Soil management: compost production and use in tropical and subtropical environments

by
H.W. Dalzell
Medak Agricultural Centre, India
and
A.J. Biddlestone
K.R. Gray
K. Thurairajan
University of Birmingham
United Kingdom

Soil Resources, Management
and Conservation Service
FAO Land and Water Development Division

Published by arrangement with the
Food and Agriculture Organization of the United Nations
by
Daya Publishing House®
A Division of
Astral International Pvt. Ltd.
New Delhi - 110 002

"This book was originally published by the Food and Agriculture Organization of the United Nations (FAO) as FAO Soils Bulletin No. 56".

First Indian Edition, 2007
Reprinted, 2016
ISBN: 978-81-7035-504-5

Published by : **Daya Publishing House®**
A Division of
Astral International Pvt. Ltd.
– ISO 9001:2008 Certified Company –
4760-61/23, Ansari Road, Darya Ganj
New Delhi-110 002
Ph. 011-43549197, 23278134
E-mail: info@astralint.com
Website: www.astralint.com

Printed at : **Chawla Offset Printers,**
Delhi-110052

PREFACE

This Soils Bulletin provides training material on composting for extension workers and teachers in countries in the tropics and subtropics. The objective is to promote the use of locally available organic materials to increase soil organic matter content for the improvement of soil fertility, and as sources of plant nutrients in conjunction with mineral fertilizers.

This manual is written for all those concerned with the maintenance and improvement of soil fertility, especially under tropical and subtropical conditions. It is hoped that it will be of value to staff in Government Agricultural Advisory Services and in non-government agencies both at the policy making level and in various levels in extension services. It contains material for use in farmer training. Those involved in planning safe waste disposal systems will also find it useful and it should stimulate thinking among elected officials in local government bodies throughout the tropics. Schoolteachers in rural areas will be able to base science lessons on it.

The manual has been written in simple language without detracting from the scientific basis or level of technology involved. This has been done deliberately to enable those whose mother tongue is not English to use it and also to form a bridge between scientific knowledge and practical compost production.

The severe drought and famine in parts of Africa in 1985 have shown the necessity for adequate soil organic matter to prevent hillside erosion and to retain moisture in the soil for crop growth. The cost of mineral fertilizers and their relative scarcity in some areas has increased the need to recycle waste organic materials as sources of crop nutrients. This Bulletin explains the basic composting process, suitable organic wastes, practical composting methods, use of the product in a variety of situations and a consideration of economic and social benefits. It also deals with approaches to practical extension work with farmers on the subject.

The book has been written by Howard Dalzell, Director of the Medak Agricultural Centre, India, in conjunction with Ken Gray, Joe Biddlestone and Kamala Thurairajan of the Compost Studies Group at the University of Birmingham, England. The latter group was formed in 1965 in the Department of Chemical Engineering and has been studying the composting process and its application to the treatment of organic wastes from garden, farm and municipal sources. During his 16 years experience of agricultural development in India, Dalzell has carried out the practice of composting with support from the university group in England. This manual is the result of their joint experiences.

It is sincerely hoped that the information given in this Soils Bulletin will provide help and encouragement to the many people involved with increasing the food supply in developing countries.

ACKNOWLEDGEMENTS

The authors are grateful to Dr. C.S. Ofori (FAO) for advice on preparing the book and to Miss F. Botts (FAO) for help with the photographs which were taken from FAO filmstrips. We thank the following for permission to use Plates 7, 8, 14, 15, 20 and 27 - Dr. G.F. Wilson and Dr. B.T. Kang, Butterworths, Faber & Faber, Farmers Weekly, Oxford University Press and the World Health Organization. Intermediate Technology Transport Ltd kindly supplied Figures 28 and 29. Rev. J.W. Barya, on rural development in Uganda, supplied information on conditions in Africa. Mr. L.D. Hills gave helpful advice on several aspects of composting. Sincere appreciation is expressed to Mrs. P. Saul for typing the manuscript. Financial assistance from several sources, particularly the Methodist Church Overseas Division in London and the Wolfson Foundation, is gratefully acknowledged. Finally, the authors thank the many enthusiastic composters, worldwide, whose knowledge has helped lay the foundations for this book.

SUMMARY

The manual combines a comprehensive description of raw materials, composting theory, all scales of composting practice and the use of the compost product in soil management. Economic and social aspects of composting are considered and attention is given to methods of extension education.

Chapter 1 outlines reasons for the widely occurring lowering of soil fertility in many parts of the tropical world which is leading to decreases in food production levels. These decreases are taking place despite the rapidly increasing food needs of expanding populations. The chapter provides an analysis of the sort of farming systems required for stable production increases and describes the way in which soil fertility is maintained under naturally occurring tropical vegetation systems. The need for composting is highlighted and a historical description is given of the worldwide development of composting.

Chapter 2 describes the microbiology, biochemistry and process factors such as aeration, moisture and temperature which must be controlled to achieve a satisfactory compost product.

Chapter 3 is a description of organic waste materials which can be composted. Consideration is given to quantities available, probable compositions and precautions to prevent loss of nutrients during storage prior to composting. The preparation of waste for composting is also described.

Chapter 4 describes practical composting processes for small heaps containing less than half a tonne of wastes and for large heaps suitable for farms and villages. Partly mechanized and fully mechanized units for composting urban wastes in towns and cities are also covered.

Chapter 5 considers the composition of the compost product and discusses the importance of soil organic matter in managing soil fertility. Attention is given to ways of using compost to maximum efficiency in kitchen gardens, market gardens, field crop farming, tree planting and fish farming. It is concluded that compost will in many cases enhance the efficiency of use of mineral fertilizers. The merits of green manuring and direct incorporation of straw and dung are described.

Chapter 6 deals with the relevance of composting to minimizing environmental pollution risks normally associated with organic wastes. These include disease-causing organisms, flies and vermin, odours, weeds, heavy metals and pollution of drinking water.

Chapter 7 examines the economic and social potential and difficulties involved in composting. Comparisons are made of mineral fertilizers and composts on a nutrient basis and costs are calculated in terms of labour. The cost of composting is considered in the context of connected factors at village level such as night soil disposal, fodder and fuel systems and the maintenance of community irrigation facilities.

Chapter 8 looks at the requirements of extension systems which aim to develop the practice of composting. Stress is laid on the need to co-ordinate the work of researchers, extension agents and farmers. Practical suggestions are given on time schedules and the training of extension workers and farmers. Possible trials to enable farmers to test the effectiveness of the compost product are outlined. These trials can be used to define any problems clearly and indicate areas for further research and extension effort.

Chapter 9 summarizes the findings of the earlier chapters and builds arguments for local bodies, agricultural and public health extension agencies and farmers throughout the tropics to take up composting. While the different agencies will have to consider different aspects of composting, there is a great underlying need for co-ordination to complement each other's work and to integrate organic recycling with fodder and fuel management.

The Appendixes provide a glossary of terms and abbreviations, notes on plant nutrient calculations, a review of the complementary nature of composting and biogas production, suggestions for further reading and information on FAO filmstrips about composting.

For those who start composting or attempt to refine their systems a wide choice of possibilities exists. The authors will be happy to deal with specific difficulties and to provide further suggestions so that existing local advisory services may be fully utilized.

CONTENTS

LIST OF TABLES

LIST OF FIGURES

LIST OF PLATES

CHAPTER 1

INTRODUCTION

The composting techniques described in this manual will be suitable for most tropical and subtropical climates found in those countries between, or close to, the Tropics of Cancer and Capricorn (approximately 30°N to 30°S of the Equator). Essentially this covers Mexico and most of South America, Africa, the Middle East, Asia and much of Australasia. It is realized that within the countries there are many areas with temperate type climates. As a general guide, precautions against loss of heat from a compost heap will be necessary where ambient temperatures fall below 10°C for extended periods, while precautions against drying-out will be required where temperatures exceed 20°C for a long period of time.

1.1 COMPOSTING

Composting is the decomposition or breakdown of organic waste materials by a mixed population of micro-organisms (microbes) in a warm, moist, aerated environment. The wastes are gathered together into a heap so that the heat which is evolved in the process can be saved. As a result the temperature of the heap rises, thereby speeding-up the basic degradation process of nature which normally occurs slowly in organic wastes which fall on to the surface of the ground. The final product of the process is compost or humus which is of value in agriculture for improving the structure and moisture-retention properties of the soil and for supplying plant nutrients as the compost finally breaks down to mineral matter. For success in carrying out composting and in using the produce to best advantage some knowledge and practical skill is required.

1.2 OBJECTIVE

This book has been written in answer to the need for information on composting organic wastes. It describes in simple terms the scientific principles involved in composting, a number of processes for composting various materials at different quantities, uses of compost and a consideration of the economics and social benefits of the practice. It should provide suitable training material for extension workers and teachers in the less developed countries who can explain the ideas to local groups, ideally in the context of 'training and visit', combining theory and practice.

The aim is to maximize the use of locally available organic waste materials in agriculture and horticulture:

i. as a source of plant nutrients which are released gradually as compost mineralizes in the soil;

ii. to increase soil organic matter content with the benefits of improved water retention, better soil workability and increased resistance to erosion.

This should improve, or at least maintain, the productivity of soils used for food production.

In recent years the importance of soil conservation and soil organic matter has been better appreciated. Organic waste recycling is highly relevant and there have been a number of conferences, study visits and workshops on this subject. The resulting books, bulletins and proceedings have been on a high technical plane and have not been easily understood at the local, village, smallholder level. This book aims to help at this level, though it also covers the situation of the more wealthy farmers owning some cattle, the owners of plantations growing cash crops and generating organic wastes, and the village, town and city community generating waste in the form of refuse, night soil or sewage sludge.

1.3 FOOD PRODUCTION IN LESS DEVELOPED COUNTRIES

The present situation of food supply in many of the less developed countries gives much cause for concern. Rapidly rising populations in Africa, S. America and S. Asia are increasing faster than the improvement in food production. Several countries which were exporters of grain in the nineteen-sixties and nineteen-seventies are now importers. The drought which gave rise to famine in the Sahel in Africa in the nineteen-seventies has given way to an even more severe drought and famine across the whole of sub-Saharan Africa in 1985 which is putting the lives of some 100 million people in peril.

The causes of the present problems are manifold. The best land, with deep fertile soils, is often used for plantations growing cash crops for export to try to service overseas debts. Such plantations are run with high efficiency using imported western technology - tractors, implements, irrigation and agrochemicals. Food production for local consumption has been pushed out into the marginal soils, often shallow, arid and on hill slopes. The problem has been made worse in many areas by extensive deforestation, to provide land for cropping and wood for local energy needs. The loss of adequate tree cover has often led to severe erosion by water and wind. In heavy rainfall, topsoil has been swept down to rivers and out to sea, leaving infertile soil or, in extreme cases, the underlying rock exposed.

Industrial development ideally needs to be based on a stable agriculture which is able to feed the population. Many of the industrialized nations are situated in temperate climates where the rate of loss of soil organic matter is relatively low and soils are fairly stable. By contrast, many of the less-developed countries are in the tropical or subtropical zones where high soil temperatures lead to far faster loss of organic matter; this immediately puts these countries at a disadvantage as their soils need a constant input of organic matter to remain productive.

An important need in increasing food production in the less developed countries is to concentrate resources on improving the productivity of the small family farm, enabling it to provide enough food, fuel and employment to reduce migration to the towns. This must start with teaching an increased respect for the soil, not regarding it as just a mass of mineral matter but as a dynamic entity, containing organic matter and living organisms, which needs protection, feeding and careful management if it is to be productive.

1.4 AGRICULTURAL SYSTEMS

In many ways each farm or smallholding is a unique enterprise. However, the extremes in systems can be taken as:

ADVANCED	SIMPLE
High cost	Low cost
High risk	Low risk
High output	Low output
High profit	Low profit
Advanced technology	Simple technology
Low labour usage	High labour usage

Advanced systems have developed in affluent countries mainly because of:

- low labour availability due to the demand for manpower by competing industries,
- a relative scarcity of land because of high population densities (though there are exceptions, e.g. Australia, Canada, USA),
- the relative stability of the soils in temperate climates,
- the ease of implementing technical advances within a well-developed infrastructure.

These developments have taken place over hundreds of years; hence social changes have not lagged behind technical advances, and in some cases have actually stimulated them.

It was originally assumed that the advanced agricultural systems of developed countries could be transferred to the less developed countries with far-reaching beneficial results. In a few cases this has been true, notably in the Indian Punjab where the 'Green Revolution' has greatly increased yields of wheat. However, there are more cases where farmers have either not adopted these techniques or have tried them and failed. The following are among the reasons for this lack of success:

Cost. A less developed country is frequently lacking in capital, particularly for imported items. High cost systems, although applicable to large cash-crop plantations, are most unlikely to find widespread application among smaller farmers, many of whom are operating at close to subsistence level.

Risk. Advanced agricultural systems with their emphasis on high output involve intensive production, such as continuous monocropping of cereals or intensive rearing of livestock. These practices are invariably accompanied by higher risks, especially from pests and diseases. The latter are normally kept within acceptable limits by improved methods of prevention or control often involving agrochemicals; these increase the cost of production but are acceptable if they succeed. By contrast, farmers in the less developed countries can afford to take very few risks, especially when in a subsistence situation; every harvest must be reasonably successful or the result is hunger. In these situations the risk can be spread, without significantly raising costs, by growing a variety of crops and practising rotations and intercropping. The major risk is then from rainfall - either too much or too little; this can best be guarded against by protection of the topsoil from water erosion using tree belts and appropriate tillage, and by maintaining a high level of organic matter in the soil to conserve soil moisture.

Output and profit. There is an urgent need in the less developed countries to increase output to relieve food shortage. At the same time small farmers need increased profits to improve their standard of living, keep their farms viable and maintain rural employment. Failure to achieve this has been one of the bad social consequences of the 'Green Revolution' even where it has been apparently successful in raising cereal outputs from the richer farmers who could afford the necessary inputs. High cost farming systems may only achieve high output at increased food prices which may not be within the means of the local economy. If the economy cannot sustain such prices then the farmer's profit margin may not rise satisfactorily.

Technology. It is not the purpose of this book to argue that technical progress is not useful or that it is needed only in the industrialized Western countries, or by rich farmers in the less developed countries. Experience has shown that advanced technology cannot be applied widely in a developing country unless the infrastructure of that country is adequate. The logistical problems encountered in getting food aid to the remote rural parts of Ethiopia, Southern Sudan and Chad in the 1985 famine have shown all too clearly the need for adequate railways, roads and spare parts to keep transport in working order. Moreover, sufficient attention must be given to agricultural training in schools and colleges and by the rural extension services so that the need for, and the means of, technical advance can be understood and interpreted in a local situation. The local economy must also be able to absorb the costs of the technical innovation prior to reaping the rewards which may only be paid back from increased profits over a period of years. The more sophisticated the technology, the less likely that these conditions can be met. For successful technological advance the need is for improved and reliable technology suitable to local conditions of labour, skills and environment.

Labour usage. The main asset of most less developed countries is a large labour pool arising from major rural unemployment. This is often viewed as a problem in terms of food supply. Clearly, in such areas, an advanced farming system which minimizes labour usage will be neither readily acceptable nor at all appropriate.

In view of the above considerations there is an urgent need in many less developed countries for farming systems of an intermediate type which comprise:

- Low cost
- Low risk
- Increased output
- Increased profit
- Modified and improved technology
- High labour usage
- Long-term stability

Such systems will be of great relevance especially in the vast areas of un-irrigated agriculture in the tropics and subtropics where farming is carried out by smallholders.

One of the most important and rewarding methods of increasing agricultural output is by raising the level of soil fertility, both by improving the long-term structural stability and moisture retention of the soil and by increasing the supply of plant nutrients. Much can be accomplished within a local community by recycling back to the soil, preferably via a compost heap, all the organic waste materials available, from crop, animal and human origins (Figure 1). If this is done successfully, there should be less need to import mineral fertilizers from outside the community, and especially from outside the country with its effect on the country's balance of payments and its requirements for an adequate transportation and distribution system.

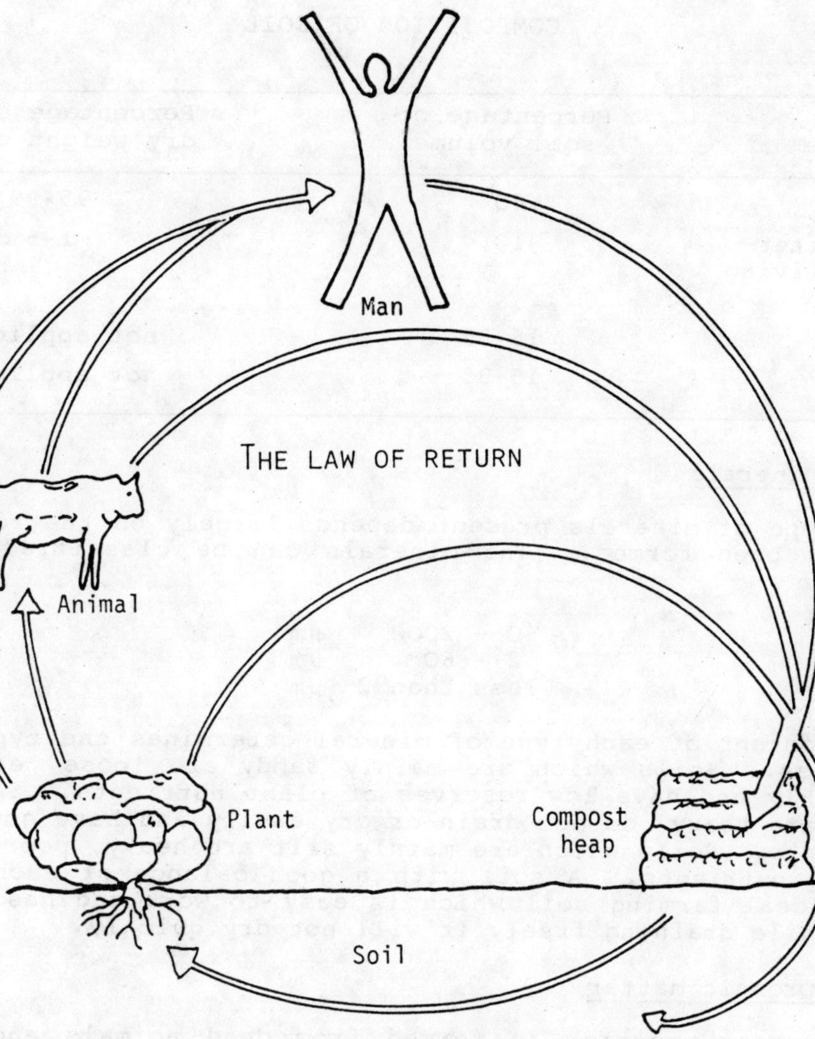

Figure 1 The cycle of life - carbon cycle

1.5 SOIL AND ITS STRUCTURE

1.5.1 Soil Formation

Soil is formed by the breakdown of rocks over millions of years. Physical breakdown occurs by the action of sun, wind, water and the roots of plants. Chemical breakdown is caused by dilute carbonic acid formed by the dissolution of carbon dioxide (CO_2) from the air into rain water and by organic acids formed when dead animals and plants decay. The broken down rock fragments are known as minerals; in addition to them the soil contains organic matter, living organisms, air and water. The approximate proportions of each soil component are given in Table 1.

Table 1 COMPOSITION OF SOIL

Component	Percentage of soil volume	Percentage of oven dry weight of soil
Minerals	40	95-99
Organic matter including living organisms	10	1-5
Air	15-35	not applicable
Water	15-35	not applicable

1.5.2 Soil Minerals

The type of minerals present depends largely on the rock from which the soil has been formed. The minerals can be classified according to their size:

Sands	60 - 2000	μm
Silts	2 - 60	μm
Clays	less than 2	μm

The content of each type of mineral determines the type of soil and its properties. Soils which are mainly sandy are loose, easily drained, dry out quickly and have low reserves of plant nutrients. Soils which are mainly clay are heavy, do not drain or dry easily and have good reserves of plant nutrients. Soils which are mainly silt are heavy, poorly drained and low in plant nutrients. A soil with a good balance of each mineral type will be an ideal farming soil which is easy to work and has good nutrient reserves. While draining freely it will not dry quickly.

1.5.3 Soil Organic Matter

Soil organic matter is formed from dead animals and plants. It always contains carbon (C), oxygen (O) and hydrogen (H) and in addition various inorganic elements such as nitrogen (N), phosphorus (P) and potassium (K). It has the ability to hold a lot of moisture and can attract up to ten times more plant nutrients than can the clay minerals. Organic matter is built up and broken down in a series of processes outlined in Figures 1 and 2. Due to the high temperatures in tropical and subtropical soils the rate of breakdown is high and it is often difficult to maintain high levels of soil organic matter. This does not mean that it can be ignored. Rather it means that greater efforts must be made to build up levels as organic matter affects almost all soil processes.

1.5.4 Living Organisms

These organisms are plants and animals varying in size from the submicroscopic viruses to insects and worms. Their importance to plant growth cannot be overestimated. Among their major activities are the following:

i. breakdown of fresh supplies of organic matter, releasing nutrients and forming humus;

ii. mixing of soil and building of soil structure. Individual soil particles are held together by the mycelial branches of fungi and actinomycetes or stuck together by bacterial gums (Figure 3);

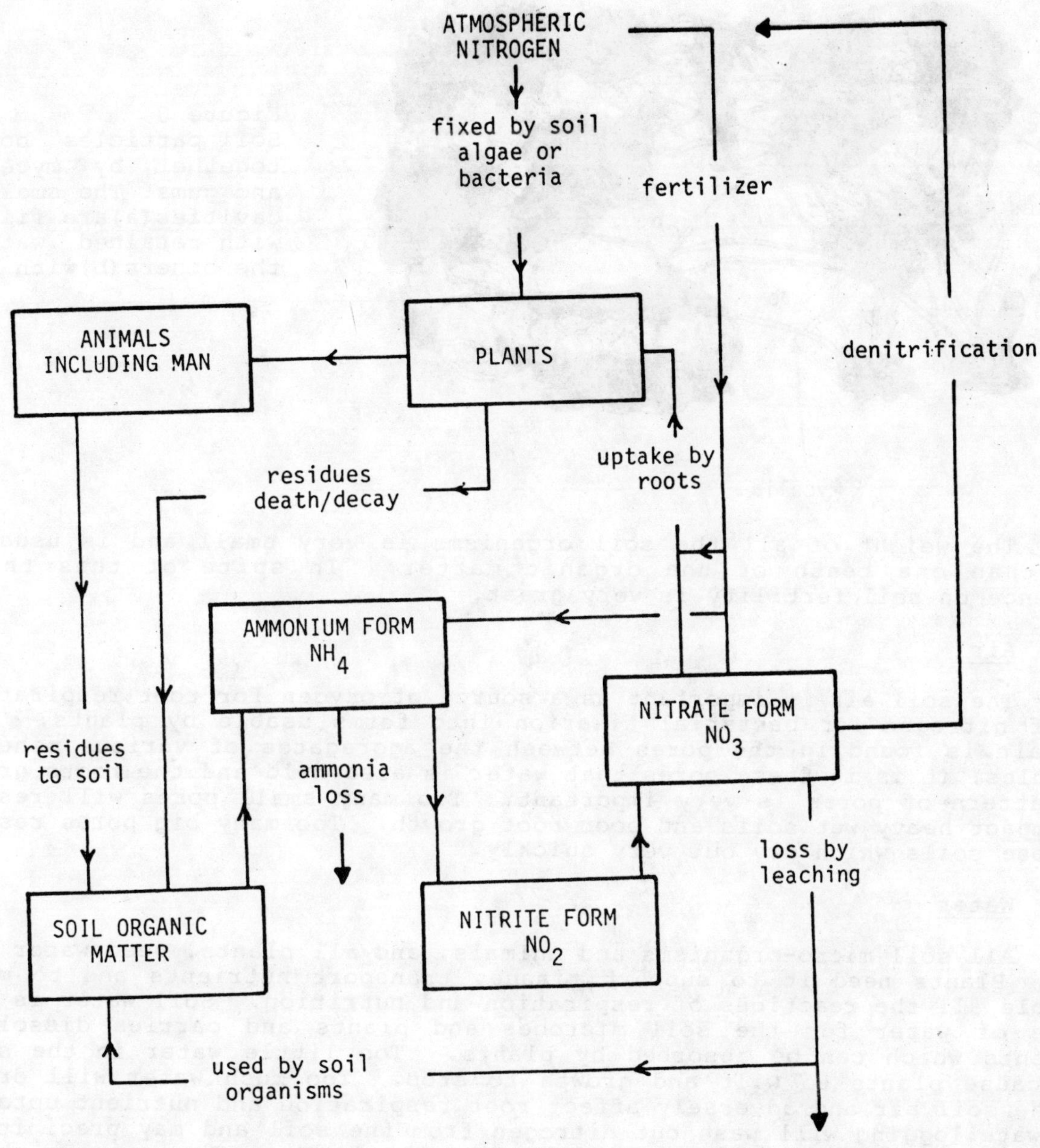

Figure 2 Nitrogen cycle

iii. creation of pores for aeration, water holding and root growth;

iv. development of intimate contacts between plant roots and soil particles which are the bridges across which nutrients pass from the soil to the plants;

v. the recovery of nitrogen from soil air into forms usable by plants. Nitrogen is one of the most important plant nutrients;

vi. the breakdown of pesticides into non-poisonous compounds.

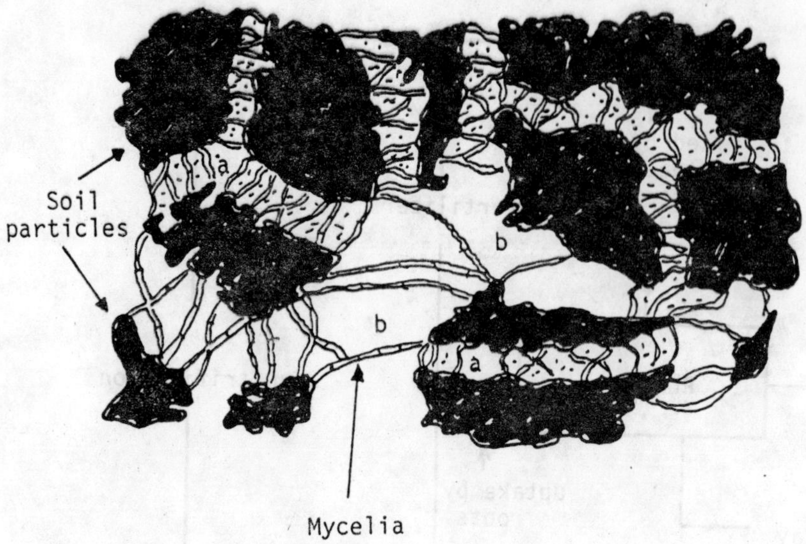

Figure 3
Soil particles bound together by mycelia and gums. The smaller cavities (a) are filled with retained water, the others (b) with air

The weight of all the soil organisms is very small and is usually less than one tenth of the organic matter. In spite of this their influence on soil fertility is very great.

1.5.5 Air

The soil air is important as a source of oxygen for root respiration and of nitrogen for bacterial fixation into forms usable by plants. The soil air is found in the pores between the aggregates of various mineral particles; it is in these pores that water is also held and the roots grow. The pattern of pores is very important. Too many small pores will result in compact heavy wet soils and poor root growth. Too many big pores result in loose soils which dry out very quickly.

1.5.6 Water

All soil micro-organisms and animals, and all plants, need water for life. Plants need it to support tissue, transport nutrients and to make possible all the reactions of respiration and nutrition. Soil water is the source of water for the soil microbes and plants and carries dissolved nutrients which can be absorbed by plants. Too little water in the soil will cause plants to wilt and growth to stop. Too much water will drive out the soil air and adversely affect root respiration and nutrient uptake. Such waterlogging will wash out nitrogen from the soil and may precipitate chemicals such as iron which become toxic to plants. The suction which plants must exert to take water out of the soil depends largely on the size of pore in which the water is held. The smaller the pore, the more difficult it is for the plant to obtain water from it.

1.5.7 Soil Structure

All the soil constituents are important in themselves but the most important thing for the growing plant is the relationship between the different constituents (Figure 3). The organic matter and the soil organisms link the mineral particles together and create the conditions in which the plant will respire, take in water and nutrients, and develop roots. Fungal branches join mineral particles and bacterial gums help to bind them into aggregates between which the air and water holding pores are formed. The minute fungal branches also grow into the plant root hairs, called mycorrhizal association, and facilitate nutrient transfer to the plants. Nutrients are released into the soil by bacterial and fungal breakdown of organic matter; this also produces humus which is a stabilized and long-lasting form of organic matter. Humus holds nutrients and moisture and stabilizes the soil, helping it to resist erosion. Bacteria largely carry out the carbon, nitrogen and other cycles which control the availability of most plant nutrients. Essentially, soil management is a process of encouraging the activity of the soil micro-organisms and animals; this can best be done by maintaining adequate levels of soil organic matter. Farmers can do very little about the mineral make up of their soil but can do a lot about the organic cycles taking place in it.

1.6 TROPICAL VEGETATION SYSTEMS AND SOIL FERTILITY

A tropical rain forest is a closed ecological system which represents a highly efficient production form (Figure 4). The tall forest trees with leaf canopy meeting overhead are balanced by a deep rooting system permeating the topsoil and subsoil down to the layer of mineral rocks which are slowly weathering. The shrub layer beneath the trees and the plants which grow on the soil surface all help to fill the above ground space with growing matter and the soil with roots. The soil is rarely exposed to the sun. A considerable quantity of organic litter falls onto the soil surface and decays (Figure 5), providing food for a highly active soil microbial and animal life; this helps with decay, with nitrogen fixation and with plant growth (by mycorrhizal association with roots and by other symbiotic relationships). The deep rooting system ensures that plant nutrients released by mineralization of organic debris, and from the weathering of the deep soil layers, are kept effectively in circulation with very little loss by leaching. In addition there is believed to be some relationship between the local microclimate and the movement of water up through the tree root system, trunk and transpiration from the leaves. Unfortunately, although a rain forest is biologically very stable, its productivity, in terms of food supply for humans, is rather low.

As the rain forest gives way to savanna a very different ecosystem results (Figure 6). The isolated trees and occasional shrubs mean that the subsoil is only rarely permeated by deep roots and the bringing up of nutrients from deep layers is localized to the immediate vicinity of the trees. Moreover the grass-herb surface layer is less efficient in storing nutrients than the more woody plants of the forest. It does, however, give the soil a degree of protection from the sun. There is now much less organic matter in the soil and nutrient leaching can occur. Productivity is lower than for the rain forest but it is relatively stable as long as the savanna is not over-grazed (Moss 1981).

Finally, when deforestation has taken place to provide land for cropping, the original biological cycle is markedly changed. The forest canopy with its production of organic debris is removed and burnt, while the root system disintegrates. There are no longer any deep roots to recycle nutrients back to the crops on the surface, nor to bring up fresh nutrients from lower strata. The supply of organic debris to the surface is greatly reduced; weed removal and seasonal cropping expose the soil surface to the sun for long periods. This exposure greatly increases the rate of oxidation of organic matter in the surface layer, often leading to a breakdown of soil structure. The loss of organic matter means that the water retention properties of the soil are seriously impaired. As a result water erosion and nutrient leaching can take place in times of heavy rainfall, while in periods of water deficiency the soil water available for crops is soon exhausted. The biological life of the soil falls to a very low level and eventually crop production is no longer possible. This is the situation in parts of sub-Saharan Africa, with the Sahara Desert steadily moving south year by year.

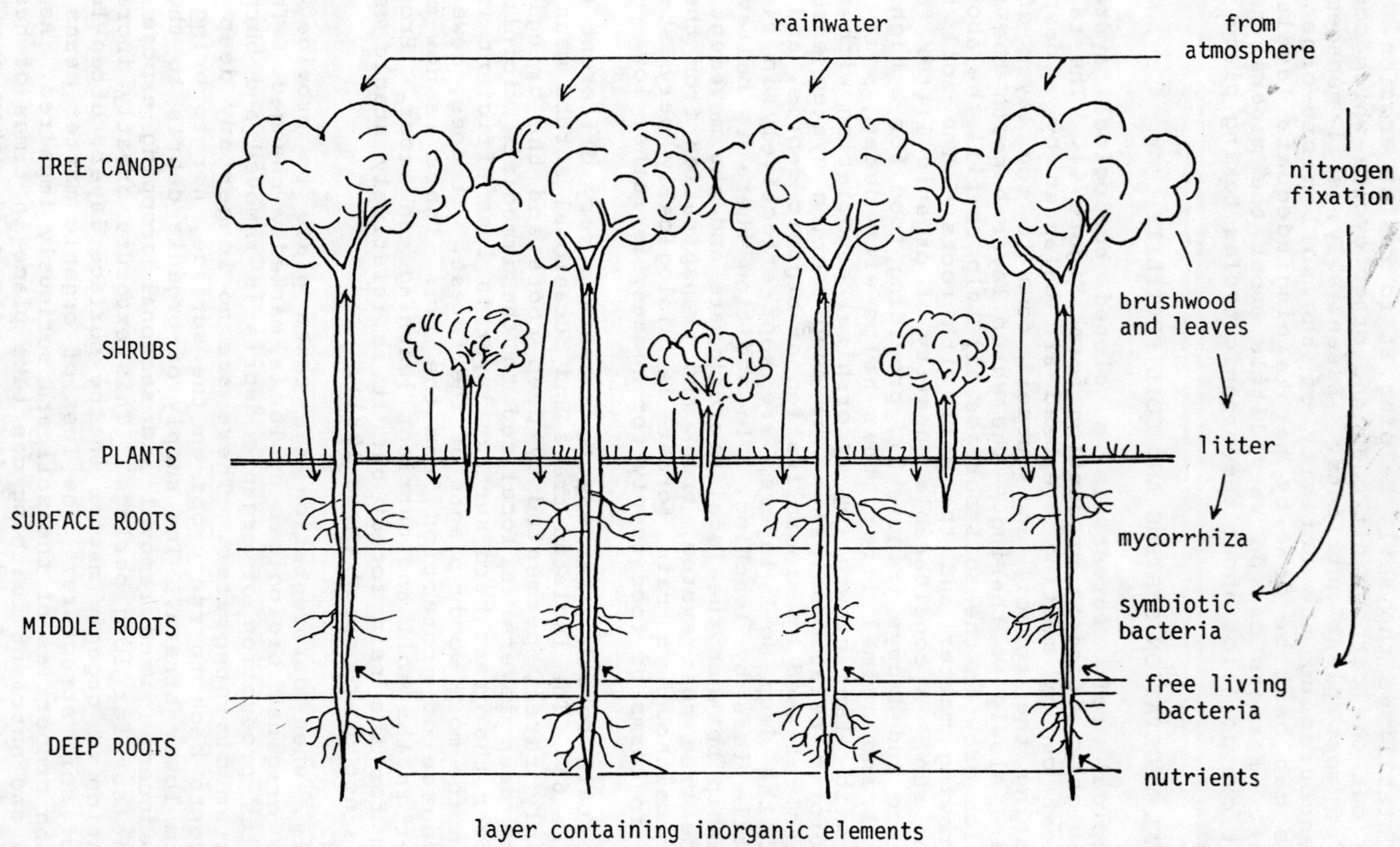

Figure 4 Diagrammatic representation of organic matter in a closed canopy tropical forest

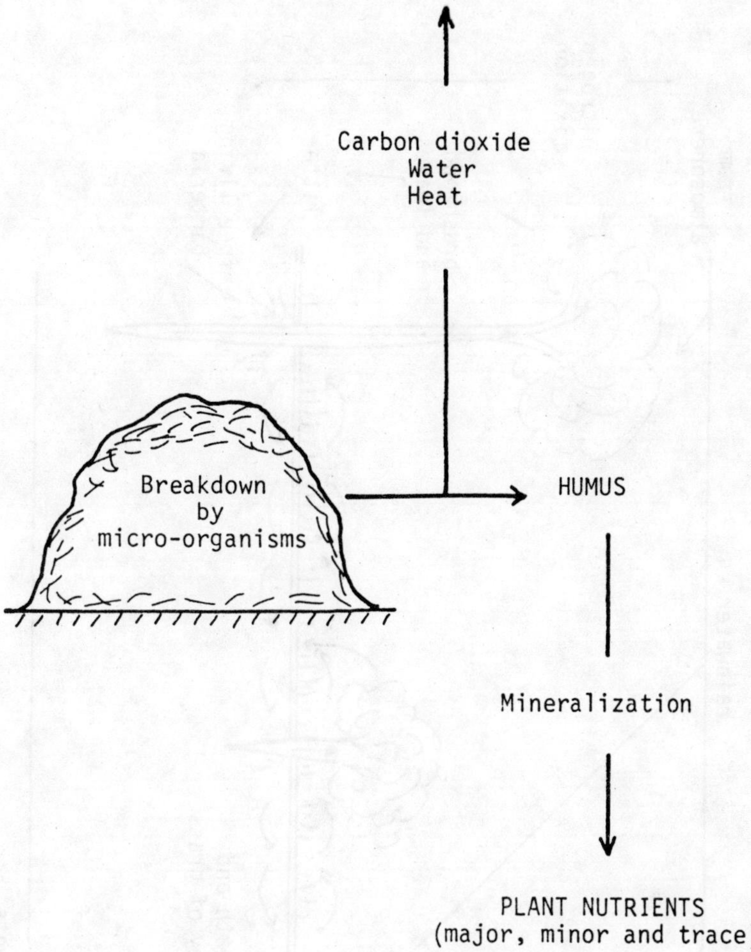

Figure 5 Organic matter breakdown

This type of situation rarely occurs in industrialized countries in temperate climates, apart from the Dust Bowl in mid-USA in the nineteen-thirties. This is because the rate of soil organic matter oxidation is much slower and there is ready access to mineral fertilizers, even though their prices rose sharply with the major increase in energy costs in the last decade. However, in these countries, more consideration is being given to a greater return of organic matter to agriculture for a variety of reasons. By contrast, the occurrence of impoverished eroded soils in many of the tropical areas is becoming far more common. Frequently such areas lack their own fertilizer industries and the cash needed to import agrochemicals from abroad.

The answer for these affected areas must lie in making better use of local resources in protecting their soils from the tropical sun and, occasionally, heavy rainfall and in increasing soil organic matter leading to conservation of plant nutrients. This must involve restoration of at least partial tree cover, soil conservation to prevent water and wind erosion, improved agricultural practices and the careful recycling of organic matter to the soil. This book deals in particular with the latter aspect.

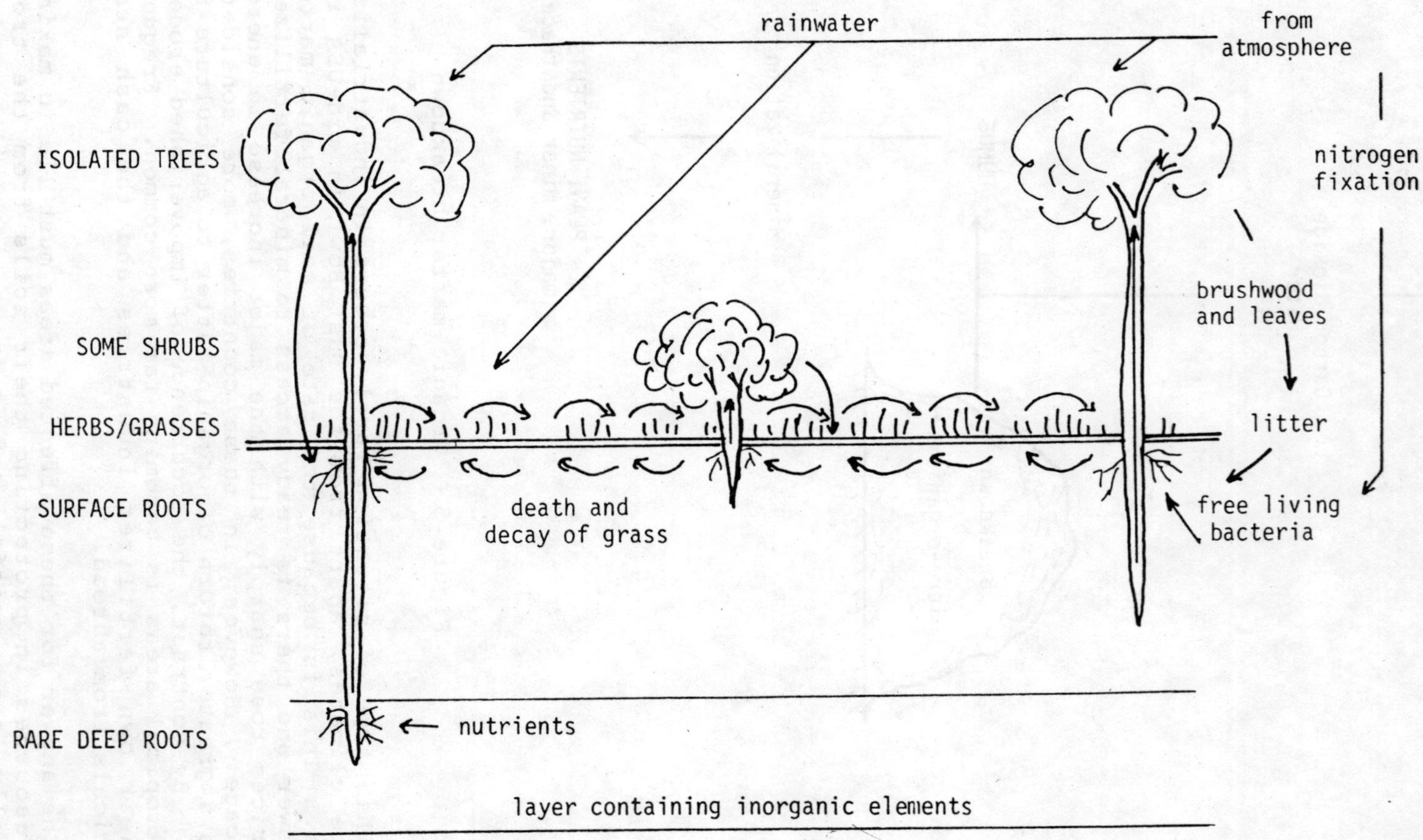

Figure 6 Diagrammatic representation of organic matter cycles in an open savanna

1.7 THE NEED FOR COMPOSTING

Organic waste materials in large quantities need to be recycled to tropical and subtropical soils in order to increase productivity. These wastes can come from many sources - crop residues, weeds, tree litter, animal manures, human wastes, suitable wastes from various industrial processes and sorted municipal wastes from towns and cities. Some of these materials can be easily incorporated directly into the soil; others will cause difficulties. All will eventually be broken down and mixed up with the soil through the work of the soil micro-organisms and tiny soil animals such as earthworms, ants and termites. However, it is usually better to put the wastes through a composting process first, even if only for a short period without waiting for full breakdown and maturity to be reached. The reasons for employing composting are as follows:

i. the final weight of mature compost is less than half that of the original wastes; the volume reduction is even greater. Hence the labour and costs for transporting and spreading in the field are much reduced.

ii. organic wastes often carry some pathogens (disease-causing organisms), of plant, animal or human origin. By exposing such material to the temperatures of 55 to 60°C normally reached within a compost heap, there is an effective kill of most pathogens, weeds and seeds so that problems are not carried forward into succeeding crops.

iii. materials such as manures, waste foodstuffs and human wastes are attractive to flies, other insects and vermin, both as sources of food and as sites for laying their eggs. Where raw wastes are spread carelessly in the field, unpleasant problems can arise. However, if the wastes are first composted under tidy, hygienic and controlled conditions then such problems can be greatly minimized. Once the material passes through its high temperature peak it becomes no longer attractive to undesirable insects and animals.

iv. in carrying out composting a number of wastes from several sources are normally blended together. This not only aids the composting process but improves the quality of the final product. At the high temperature of the compost heap, organic wastes break down much faster than at ambient temperature.

v. by composting the wastes first the product becomes friable or crumbly, is much more easy to handle and spread around plants or incorporate into the soil, and vastly reduces the disease risk from handling the material.

vi. soils should be used for growing crops, not as environments for breaking down organic wastes, especially where increased food production is vital. Accordingly, wastes should be broken down in a compost heap first. After a reasonable degree of maturity has been reached, the compost can be incorporated into the soil without causing a harmful effect on crop growth. The compost then comes quickly into equilibrium with the soil. By contrast, the breakdown of large quantities of organic wastes in the soil is certain to interfere with the growth of many crops. During the initial stages of breakdown there is a need for oxygen and possibly nitrogen; simple organic acids and ammonia (NH_3) are formed. A period of major disruption to the soil processes can occur and seriously affect, if not prevent, the germination of seeds and the growth of young seedlings. Established trees and shrubs can be mulched satisfactorily with partly-matured material which has only been subjected to the high temperature composting stage for pathogen control; however, most seeds and seedlings need mature compost.

Along with these advantages composting does have some disadvantages. It is an extra operation within the farming year which uses labour for collection and preparation of the wastes, building the heap and turning. It requires organizational ability and some practical skills. However, with forethought and planning composting can be arranged to take place during periods of the farming year when labour requirements are otherwise low. Secondly, if the composting process is badly carried out, there can be a significant loss of nitrogen as ammonia; in addition, if the heap is inadequately protected during periods of heavy rain, water-soluble nutrients can be leached out of the heap into the soil. Such a loss of nutrients is undesirable but can be minimized if care is taken.

1.8 HISTORICAL BACKGROUND OF COMPOSTING

The composting of organic wastes has been practised to some extent for centuries by farmers and gardeners in many parts of the world. Probably the outstanding example has been that of the Chinese in the river deltas who, by returning to the soil their crop residues, human wastes and alluvial mud swept down the rivers and canals, have been able to support high population densities. By practising excellent horticulture, with high inputs of human labour, the land has remained productive for some 4 000 years without the soil fertility and structure giving way under such pressure. Other noted exponents of composting are the people of the Hunza Valley in the Himalayas who have practised their agriculture in terraced fields on the mountain side.

Composting, as practised by the Chinese, has probably changed very little over the centuries, being essentially a small-scale batch operation. With the adoption of the process by the Western world during this century, some progress has been made in the understanding of the fundamental reaction and its application to large-scale and continuous waste treatment.

The arousal of interest in composting in the West probably stemmed from an extended visit to China, Japan and Korea by Professor F.H. King of the U.S. Department of Agriculture in the early nineteen-hundreds; his observations were carefully recorded. His book was read by Sir Albert Howard, a British economic botanist employed by the Indian Government, who was able to put King's observations on composting in China to the test in an Indian context at Indore. After several years of experimentation in the nineteen-twenties, Howard established that his Indore method of composting gave good results in terms of the vegetable and animal wastes, the supply of labour and the climatic conditions available in his district. A large number of chemical analyses were made at Indore; these indicated that in a well-constructed compost heap there was normally a net gain in nitrogen and that mixed wastes composted better than single materials.

In the nineteen-thirties the Indore method was taken up by plantation industries, farms and gardens in many parts of the world. As a result, a number of minor changes in procedure were incorporated which increased the output per man employed. There is little doubt that Howard and his team aroused much interest in the use and further study of compost. Over the same period, the field-scale investigations of Howard were being complemented by the essentially laboratory-scale studies of Waksman and his team in the USA on humus and the interplay of the micro-organisms involved in its production from organic wastes. A little earlier Hutchinson and Richards in the UK worked on the production of 'artificial farmyard manure' by composting straw with the activator 'Adco' as a chemical source of the nitrogen needed in the reaction. As a result of all this research, by the nineteen-thirties some appreciation of the effect on composting of the major physical and chemical parameters had been gained, together with some knowledge of the microbial interactions involved.

The more recent mechanized approach to composting has arisen in response to the need for continuous, controlled and hygienic disposal processes to deal with the vast quantities of municipal organic wastes arising in towns and cities from Man's exploding population and rising standards of living. Over the past 40 years at least 30 different processing schemes have been tried out for composting refuse and/or sewage sludge, with varying degrees of success. Equipment for feed preparation and product finishing is very similar in most of these processes. The major differences have been in the actual composting or fermentation section which has been attempted in a wide variety of pits, heaps, cells, bins, digestors, silos and rotating drums. A good review of the state of the art in the nineteen-fifties is given by Gotaas (1956). At that time few processes had a capacity of over 100 tonnes/day; nowadays 500 t/d units are being installed and 1 000 t/d ones are being considered. Some systems still employ pulverization followed by fermentation in open elongated heaps (windrows); an appreciable number use the Dano rotating drum followed by maturing in windrows; a few are starting to employ a high degree of automation in vertical, multi-floored silos with continuous agitation and control of aeration and moisture (Biddlestone *et al*. 1981). A major problem with these processes is that they are handling large volumes of waste materials and producing a product which is both bulky and has rather a low market price. Accordingly expensive treatment operations can rarely be economically justified. After an initial upsurge of interest in the ninteen-fifties surprisingly few further refuse composting units have been installed in the West. Most interest has been shown in the Near East in the oil-rich states where finance is more readily available and there is a good demand for compost for reclaiming desert soils. Very recently nearly 100 units have been built or designed in the USA and France for the composting of sewage sludge with a 'bulking agent' such as wood chips, sawdust or fragmented rubber tyres.

Following the sharp increase in oil prices in the nineteen-seventies and their effect on fertilizer costs it became clear that the agricultural systems of many less developed countries could not be based on an assured input of mineral fertilizers. Accordingly, increasing emphasis has been placed on trying to supply plant nutrients by the recycling of organic wastes. Several conferences, workshops and study visits (particularly to China) have been held and the resulting information published (FAO 1975, 1978a, 1978b, 1978c, 1980, 1982). The need for this new approach has been dramatically highlighted by the 1985 drought and famine in sub-Saharan Africa. This has shown that for many areas with soil problems, the high technology agriculture of the West has little relevance.

The only way forward at present must come from greater concern for the soil, improved husbandry and increased recycling of organic wastes back to crop growing on millions of smallholdings. A significant part in this technology will be played by the compost heap.

CHAPTER 2

PRINCIPLES OF COMPOSTING

2.1 GENERAL

Composting is the breakdown of organic material by large numbers of micro-organisms in a moist, warm, aerated environment to give a humus end product. The micro-organisms are the very tiniest and simplest members of the plant and animal kingdoms.

An outline of the composting process is given in Figure 7. The micro-organisms take in moisture, oxygen from the air and food from the organic material. The organisms give off carbon dioxide, water and energy; they reproduce themselves and eventually die. Some of the energy released is used for growth and movement, the remainder is given off as heat. As a result a heap of composting material passes through warming-up, peak temperature, cooling down and maturing stages.

The process is most commonly used to change waste organic materials into a useful product. The waste material will normally have within it a variety of micro-organisms capable of carrying out the process. When the material is exposed to the air and the moisture content brought to a suitable level the organisms begin their work. In addition to oxygen from the air and moisture, the micro-organisms require for their growth and reproduction a supply of food containing carbon and nutrients such as nitrogen, phosphorus and potassium. These food requirements are usually provided by the waste materials.

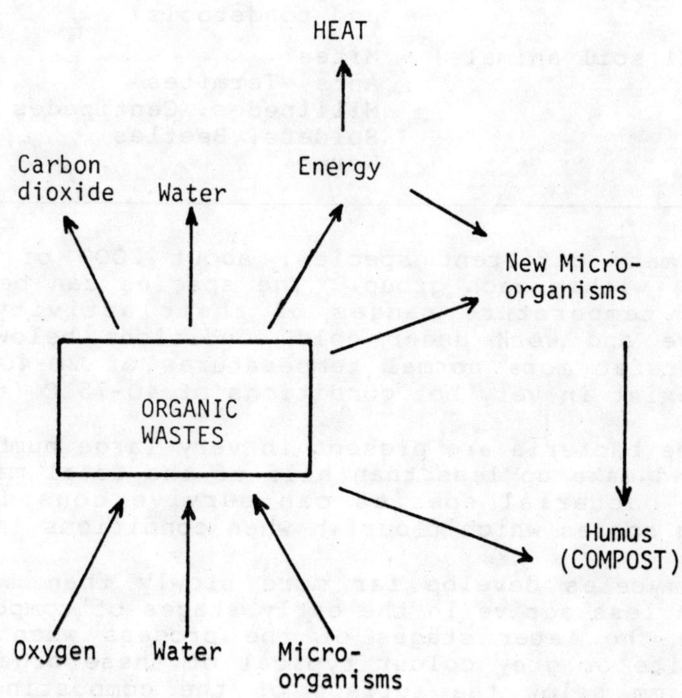

Figure 7 The composting process

The final compost product is made up of the more resistant parts of the original organic matter, products from the breakdown process, dead micro-organisms and some living micro-organisms, together with products from further reaction between these materials. For the process to give a satisfactory compost product the organisms must be given optimum conditions of food, air, moisture and warmth.

2.2 MICROBIOLOGY

Composting is a constantly changing microbial process brought about by the activities of a succession of various groups of micro-organisms, each of which is appropriate to an environment of relatively limited duration. A list of the main classes of organisms involved in the composting process is given in Table 2 and the organisms are illustrated in Figure 8. These organisms represent both the plant and animal kingdoms.

Table 2 ORGANISMS IN COMPOSTING

Organism type		Numbers per g of moist compost
Microflora (very small plants)	Bacteria	10^8-10^9
	Actinomycetes	10^5-10^8
	Fungi, moulds, yeasts	10^4-10^6
Microfauna (very small animals)	Protozoa	10^4-10^5
Macroflora (larger plants)	Fungi (mushrooms and toadstools)	
Macrofauna (small soil animals)	Mites	
	Ants, Termites	
	Millipedes, Centipedes	
	Spiders, Beetles	
	Worms	

There are many different species, about 2000 of bacteria and at least 50 of fungi, within each group. The species can be further divided according to the temperature ranges of their activity. Some of the organisms can live and work under cold conditions below 20°C (psychrophiles), others prefer more normal temperatures of 20-40°C (mesophiles), whilst a few can exist in very hot conditions of 40-75°C (thermophiles).

Although the bacteria are present in very large numbers, they are of very small size and make up less than half of the total mass of the micro-organisms. Some bacterial species can survive considerable heat and dryness by forming spores which flourish when conditions improve.

The actinomycetes develop far more slowly than most bacteria and fungi and are much less active in the early stages of composting. They are more prominent in the later stages of the process when they can become abundant. The white or grey colour typical of these organisms is clearly visible some 100 mm below the surface of the composting mass (Bertoldi 1983).

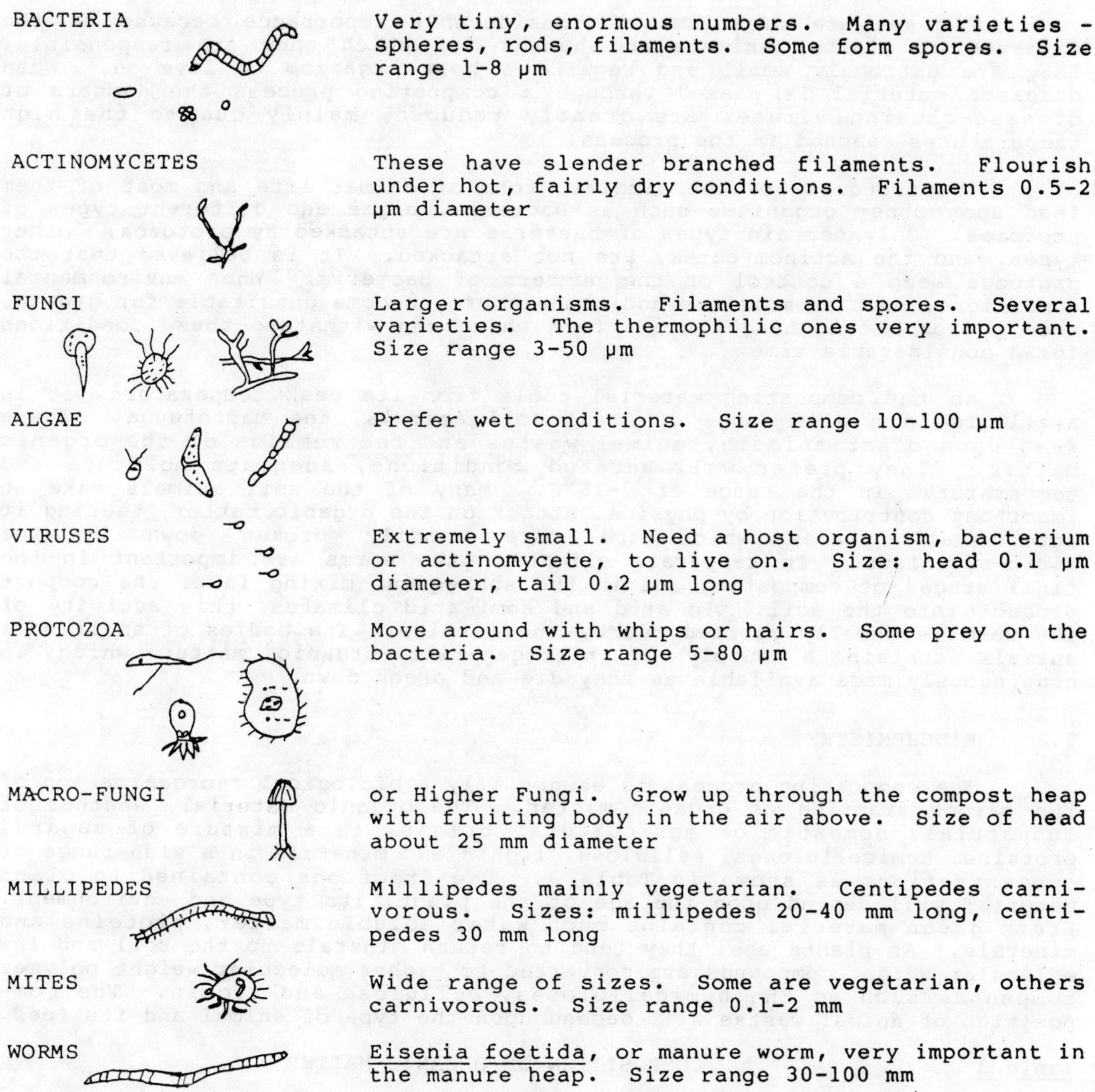

Sizes: 1 μm (1 micron) = 10^{-6}m = 0.001 mm

Figure 8 Organisms in compost heaps

Fungi are very important in the breakdown of cellulose which is a more resistant part of the organic matter and can form up to 60 percent of the total mass of material. The conditions within composting masses should be arranged to encourage the activities of these fungi. Temperature is an important consideration as the fungi will die out as the temperature rises above 55-60°C, reinvading from cooler zones as the temperature falls.

The algae require sunlight to carry out their work and prefer wet conditions. In the composting process their numbers are small and they are not very significant.

Viruses are organisms of considerable importance because of the diseases of plants, animals and humans for which they are responsible. They are extremely small and require a host organism to live on. When diseased material is passed through a composting process the numbers of disease-causing viruses are greatly reduced, mainly due to the high temperatures reached in the process.

The protozoa are the simplest form of animal life and most of them feed upon other organisms such as bacteria, algae and different types of protozoa. Only certain types of bacteria are attacked by protozoa; other types, and the actinomycetes, are not attacked. It is believed that the protozoa keep a control on the numbers of bacteria. When environmental conditions, such as moisture and temperature, become unsuitable for growth, then protozoa can change into a form which can withstand these conditions for a considerable time.

As the composting material cools from its peak temperature, it is available to a wide range of small soil animals, the macrofauna. These feed upon other animals, animal wastes and the remains of the organic matter. They prefer well aerated conditions, adequate moisture and temperatures in the range of 7-15°C. Many of the soil animals make an important contribution by physical attack on the organic matter, tearing it into small pieces which are more readily broken down by the micro-organisms. In temperate climates, the worms are important in the final stages of composting and in the subsequent mixing in of the compost product into the soil. In arid and semi-arid climates, this activity of the worm is usually carried out by the termite. The bodies of small soil animals contain a supply of nitrogen-rich organic matter which is continuously made available as they die and break down.

2.3 BIOCHEMISTRY

The composting process is essentially a biological reorganization of the carbon fraction of organic matter. The organic material, whether of industrial, domestic or agricultural origin, is a mixture of sugars, proteins, hemicelluloses, cellulose, lignin and minerals in a wide range of concentrations as shown in Table 3. The fractions contained in plant material will depend upon the age of the plant, its type and environment. Fresh green material contains much water soluble matter, proteins and minerals. As plants age, they tend to return minerals to the soil and low molecular weight compounds are converted to higher molecular weight polymer compounds such as the hemicelluloses, cellulose and lignin. The composition of animal wastes will depend upon the type of animal and its feed.

Table 3 COMPOSITION OF ORGANIC MATTER

Fraction	% in dry weight
Sugars, starches, amino acids, urea, ammonium salts (hot/cold water solubles)	2 to 30
Fats, oils, waxes (ether/alcohol solubles)	1 to 15
Proteins	5 to 40
Hemicelluloses	10 to 30
Cellulose	15 to 60
Lignin	5 to 30
Mineral matter (ash)	5 to 30

Composting is both a building up process and a breaking down process. The cell wall of the micro-organism attacking the organic matter is an important barrier to the movement of materials from outside the cell to inside, and vice versa, Figure 9. Some organic compounds, such as sugars, are simple in form and easily soluble in water; these can be readily taken in by the micro-organisms, providing energy and being built up into polymers. Other substances, such as the hemicelluloses and cellulose, have large molecules and must first be chopped up by enzymes (biological catalysts) before they can be used. Lignin is a woody material, which is highly resistant and only breaks down after a very long period of time. An outline of the reorganization of the different fractions - solubles, polymers and resistibles - by the cells of the micro-organisms is shown in Figure 9.

It can be seen therefore that during the composting process some of the organic material is broken down because it us used by the micro-organisms as a food source. It is found from experience that up to 50 percent of the dry weight of the starting material will be lost and that most of this can be accounted for as loss in hemicelluloses and cellulose which form the bulk of the original waste material. The fungal population is very important in the breakdown of these higher polymer materials. The bodies of the micro-organisms, both living and dead, form an important part of the compost product.

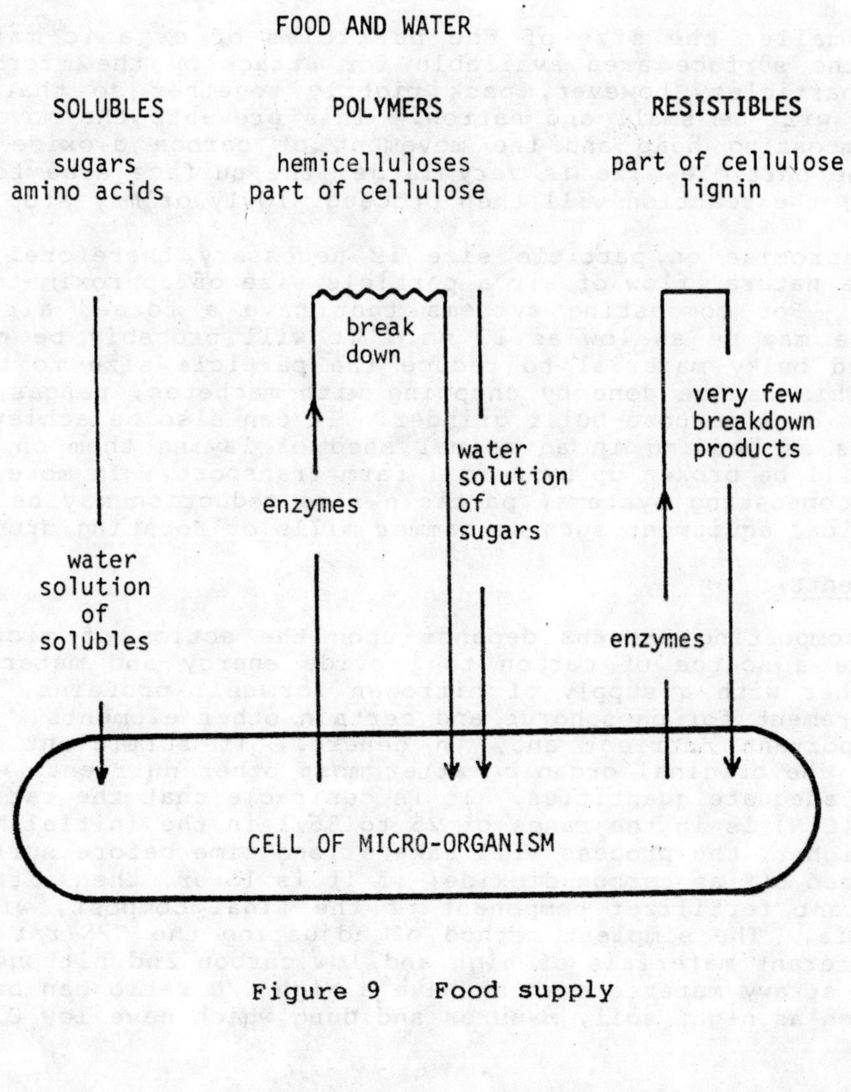

Figure 9 Food supply

2.4 PROCESS FACTORS

2.4.1 General

The breakdown of organic matter during composting is a constantly changing situation in which temperature, pH and food availability vary. The numbers and species of organisms also change during the process. The rate of progress towards the mature end product is dependent upon several process factors. These include nutrient supply, particle size, moisture content, structural strength, aeration, agitation, heap acidity (pH) and size of heap. It is desirable to provide the most suitable operating conditions for a given situation; these usually depend on labour availability and cost.

2.4.2 Separation

Good quality compost contains a high organic matter content and a minimum of non-organic material. Some compostable wastes, particularly from industrial areas, can contain high levels of metals such as copper, lead, nickel and zinc. Other non-organic materials such as glass, plastics and man-made fibres should also be removed if possible. On a small scale such separation may be done manually. On a large scale machines may be used.

2.4.3 Particle Size

The smaller the size of the particles of organic material, the greater is the surface area available for attack by the micro-organisms. Very small particles, however, pack tightly together so that the spaces between them will be small and narrow. This prevents the movement of air into the composting heap and the movement of carbon dioxide out of the heap. If the particle size is very large, the surface area for attack is much reduced; the reaction will then proceed slowly or may stop altogether.

A compromise on particle size is necessary therefore. For heaps that employ a natural flow of air a particle size of approximately 50 mm is appropriate. For composting systems that have a forced air supply the particle size may be as low as 10 mm. It will probably be necessary to chop or shred bulky material to reduce the particle size to the range of 10-50 mm. This can be done by chopping with machetes, pangas or sickles, chaff cutters or a purpose-built grinder. It can also be achieved by using the materials as bedding in an animal shed or laying them on a farm road where they will be broken up by normal farm transport. In more complicated large-scale composting systems, particle size reduction may be achieved by using mechanical equipment such as hammer mills or rotating drums.

2.4.4 Nutrients

The composting process depends upon the action of micro-organisms which require a source of carbon to provide energy and material for new cells, together with a supply of nitrogen for cell proteins. There is a lesser requirement for phosphorus and certain other elements. Nitrogen is the most important nutrient and, in general, if sufficient nitrogen is available in the original organic matter most other nutrients will also be available in adequate quantities. It is desirable that the ratio of carbon to nitrogen (C/N) is in the range of 25 to 35/1 in the initial mixture. If it is much higher, the process will take a long time before sufficient carbon is oxidized off as carbon dioxide; if it is lower, then nitrogen, which is an important fertilizer component of the final compost, will be given off as ammonia. The simplest method of adjusting the C/N ratio is to mix together different materials of high and low carbon and nitrogen contents. For example, strawy materials which have a high C/N ratio can be mixed with materials such as night soil, manures and dung which have low C/N ratios.

In tropical situations fresh green wastes such as weeds are frequently used for forage while dung is sometimes burnt for fuel. As a result, the wastes available for composting tend to be those with rather high C/N ratios. It is important, therefore, to conserve liquid manure from animal housing as this supplies nitrogen; moreover, it also contains potassium and trace elements. In many cases facilities are not provided for the collection of liquid manure; in these situations the soil or bedding under the animals can be collected from time to time and used for composting. Nitrogen may also be added to compost heaps in the form of organic fertilizers such as bone meal, hoof and horn meal, oil cakes and dried blood. Inorganic nitrogen fertilizers such as urea and ammonium nitrate may also be used. A list is given later in Chapter 3 (Table 5) of some organic materials suitable for composting, together with their nitrogen contents and C/N ratios.

Phosphorus is a less important nutrient in composting than nitrogen but is sometimes deliberately added. There is some evidence that the loss of nitrogen as ammonia from composting heaps of low C/N ratio may be partially reduced by the addition of extra phosphorus-containing material; this may not be practicable because of the cost. Recent work has shown that in a compost heap rock phosphate is changed from a water-insoluble form to one which is more available for plant use.

In order to maximize the nutrient content of the product compost, it is important to reduce serious leaching from the heap by protecting it against heavy rain and waterlogging.

2.4.5 Moisture

All organisms require water for life. At moisture contents below 30 percent on a fresh weight basis the biological reactions in a compost heap slow down considerably. At too high a water content the spaces between the particles of the material become waterlogged, preventing movement of air within the heap. The optimum moisture content of ingredients for composting is 50-60 percent. The maximum practical moisture content depends on the structural wet strength of the materials. Weak materials, such as paper, collapse readily on composting, the pores fill with water and anaerobic conditions set in. Stiff materials, such as straw and twigs, retain their wet strength for a long time and can be composted at high moisture contents. For practical purposes the material should be as damp as a squeezed-out sponge.

Water is produced during the composting process by the micro-organisms and is lost by evaporation into the air stream. In some composting processes where forced air blowing is used, moisture loss can be excessive; this may also be the case with natural aeration in very hot climates. It may be necessary therefore to provide additional moisture to the compost heap. This can be supplied as water or as further materials of high moisture content such as citrus fruit and water melon wastes.

In tropical conditions the materials will often be drier than in temperate climates; they will also dry out more quickly during composting. Care must be taken to ensure an adequate moisture content at all times. This can be achieved in practice by the following:

i. wetting the mixture initially, and when convenient during the process;

ii. composting in pits during hot weather to reduce evaporation. The use of pits during a monsoon period may lead to waterlogging; in this season the heap should be built above the ground;

iii. making heaps in the lee of buildings or trees, or the use of artificial wind-breaks;

iv. shading of the heap from direct sunlight;

v. siting heaps or pits with their long sides at right angles to the direction of the prevailing wind.

In dry tropical conditions water may be a limiting factor to crop growth. In a marginal water situation it is questionable whether to use water for composting or for irrigation. The production of 1 000 kg of finished compost may require up to 2 500 litres of water. A dressing of 25 t/ha of compost would thus require 62 500 litres/ha which is equivalent to an irrigation of 6.25 mm.

Many factors have to be considered in assessing whether such an irrigation would be of greater benefit than a dressing of 25 t/ha of compost. Among these factors are the loss of water by seepage and evaporation before the irrigation reaches the growing plant, the moisture status of the soil at the time of irrigation, the effect of compost on the moisture retention characteristics of the soil, the nutrients supplied by the compost, the soil type and the water management practised. On balance, and taking a long-term view of soil management rather than considering just a single crop, it is considered that up to a dressing of 25 t/ha compost production will represent a better usage of water.

2.4.6 Aeration

An adequate supply of air to all parts of a compost heap is essential in order to supply oxygen for the organisms and to flush out the carbon dioxide produced. Absence of air (anaerobic conditions) will lead to different types of micro-organisms developing, causing either acidic preservation (similar to silage making) or putrefaction of the heap producing bad odours.

Aeration is achieved by natural movement of air into the compost heap, by turning the material over regularly by hand or with a machine, or by blowing air into the heap using a fan. Natural movement of air makes use of the 'chimney effect' with the warm convection currents rising up through and out of the heap. However, this can fail to supply adequate oxygen in the early stages of the process, leading to anaerobic conditions in the lower central regions of the heap of material. No part of the composting mass should be more than 750 mm from an unrestricted source of air. This can be achieved in practice by laying brushwood beneath the heap and incorporating vertical air vents in the heap at the time of building and turning.

It should also be realised that a flow of air not only removes carbon dioxide and water produced in the breakdown reaction but also removes heat by evaporating moisture. The removal of heat in this way can be very important in large-scale aerated composting processes.

2.4.7 Agitation

In composting systems that rely upon natural air flow the lower central regions of the heap may be short of oxygen because the amount of air moving into the heap is inadequate. In such cases turning the material by hand or by machine allows air to reach these areas. Agitation also helps to break up larger pieces of material, exposing fresh surfaces to attack by the organisms. Control of the agitation of the heap ensures that most of the material is subjected to the highest temperature reached. However, too much agitation can lead to excessive cooling and drying of the composting material. Turning heaps of material can be expensive in labour or machine costs and the number of turns carried out is a compromise between economics and the requirements of the process. In simple composting systems using natural air flow, turning the heap 2 or 3 times should be sufficient. Such turnings also provide an opportunity to add extra moisture to the heap.

Agitation in mechanized composting systems is usually achieved by means of slowly rotating drums or specially designed arms moving through the material, turning and mixing. Too much agitation may also destroy the physical structure of the materials, causing a wet soggy mass through which air cannot pass. For instance when water melon wastes are being composted agitation may have to be reduced to avoid this problem.

2.4.8 Temperature

When organic material is gathered together for composting some of the energy released by the breakdown of the material is given off as heat and this causes a rise in temperature. The normal temperature-time curve of a compost heap is shown in Figure 10. This shows that the heap passes

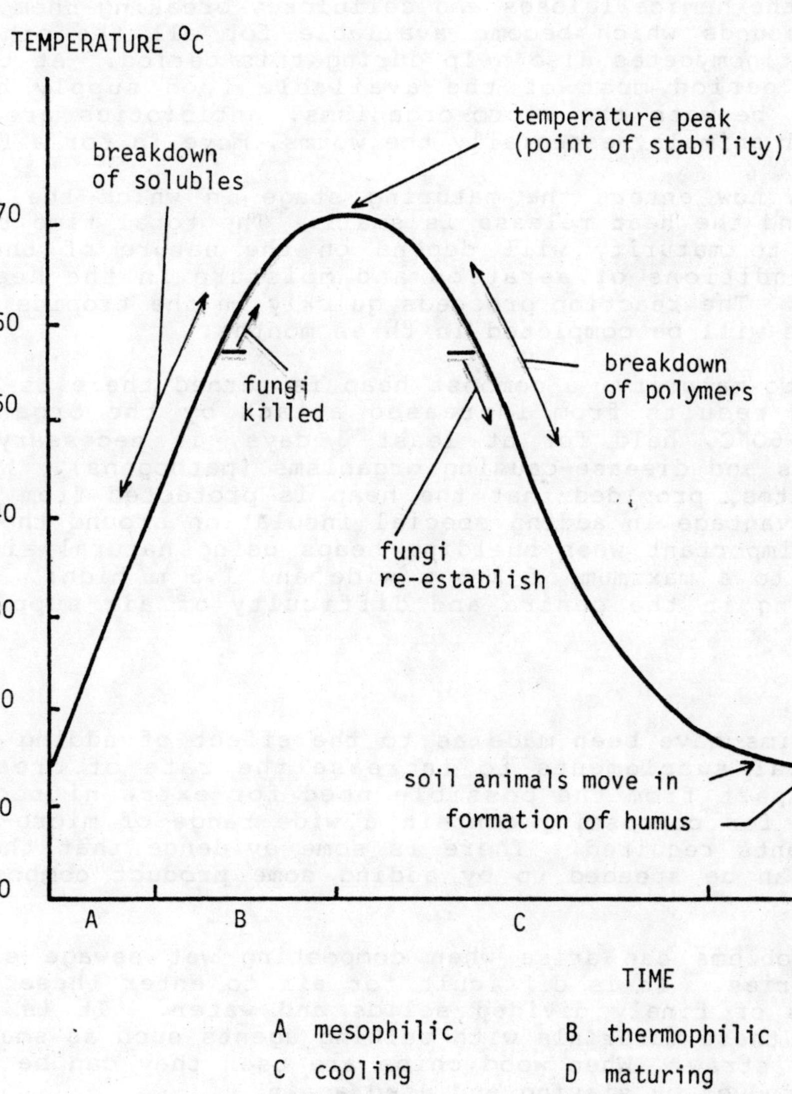

Figure 10 Temperature variations in a compost heap

through the stages of warming up, peak temperature, cooling down and maturing. At the beginning of the process the material is at ambient temperature. In the first stage, warming up, the micro-organisms present on the material multiply rapidly and the temperature rises. During this period all the very reactive compounds such as sugars, starches and fats, are broken down. When the temperature reaches 60°C the fungi stop working and the breakdown is continued by the actinomycetes and the spore-forming strains of bacteria. The breakdown slows and the temperature peak is reached. At the peak temperature the heap is losing as much heat as the micro-organisms produce.

When the composting material has passed through the temperature peak, the heap reaches stability at which the easily converted materials have been broken down and most of the high level of oxygen demand has been met. The material is no longer attractive to flies and vermin and should not give off bad odours.

When cooling down, the straws and stalks are decomposed, mainly by the fungi. This is because as the temperature falls below 60°C the fungi reinvade from the cooler regions of the heap and attack the less reactive compounds such as the hemicelluloses and cellulose, breaking them down into simpler sugar compounds which become available for all the other micro-organisms. The actinomycetes also help during this period. At the end of the cooling down period most of the available food supply has gone, competition starts between the micro-organisms, antibiotics are released and the larger soil animals, especially the worms, move in for a few weeks.

The process now enters the maturing stage in which the amount of breakdown is low and the heat release is small. The total time taken from heap construction to maturity will depend on the nature of the organic materials, the conditions of aeration and moisture in the heap and on ambient conditions. The reaction proceeds quickly in the tropics and under the best conditions will be completed in three months.

If, prior to maturity, a compost heap is turned there is a rise in temperature which results from increased attack by the organisms. A temperature of 55-60°C, held for at least 3 days, is necessary to kill virtually all weeds and disease-causing organisms (pathogens). With large heaps in hot climates, provided that the heap is protected from the wind, there is little advantage in adding special insulation around the composting mass. It is important when building heaps using natural air flow to restrict the size to a maximum of 2.5 m wide and 1.5 m high; this will prevent overheating in the centre and difficulty of air supply to the interior.

2.4.9 Additives

Various claims have been made as to the effect of adding chemical, herbal or bacterial supplements to increase the rate of breakdown in compost heaps. Apart from the possible need for extra nitrogen, most materials suitable for composting contain a wide range of micro-organisms and all the nutrients required. There is some evidence that the initial temperature rise can be speeded up by adding some product compost from a previous heap.

Aeration problems can arise when composting wet sewage sludge and animal manure slurries. It is difficult for air to enter these materials which are mixtures of finely divided solids and water. It is necessary therefore to blend these materials with bulking agents such as small pieces of wood (chips) or straw. When wood chips are used they can be separated from the compost product by sieving and used again.

2.4.10 <u>Heap Reaction, Acidity or pH</u>

The usual pattern of changes of acidity or pH in the composting process is illustrated in Figure 11. This shows that the material becomes slightly acidic at the start of composting as the initial products of breakdown are the simple organic acids. The heap then turns slightly alkaline after a few days as proteins are attacked and ammonia is released. Highly alkaline conditions will lead to excessive loss of nitrogen as ammonia; accordingly it is wise not to add lime to a heap. Highly acid initial conditions may lead to a failure of the heap to warm up. If careful attention is paid to the mixing of materials, moisture content and aeration, there is no necessity to influence the pH of the process.

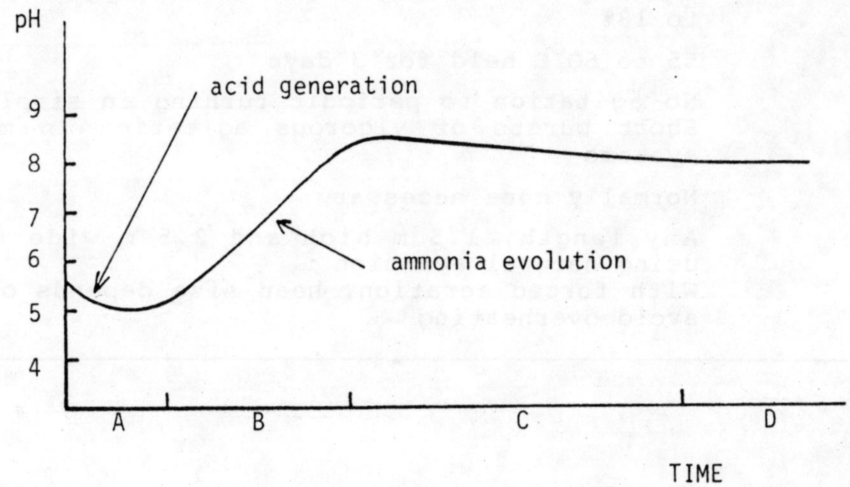

A mesophilic B thermophilic

C cooling D maturing

Figure 11 pH variation in a compost heap

The amount of ammonia lost from a compost heap can be reduced by adding a little soil, about 1% of the weight of the heap, well mixed in with the other ingredients. In some heap systems where bamboo poles are used to make air channels the heap is covered with mud; this helps to reduce ammonia loss.

2.4.11 <u>Optimum Process Conditions</u>

A summary of the most desirable values of the important process factors is given in Table 4.

The aim is to apply these factors in low cost but reliable composting systems. The complexity of the composting equipment and the degree of approach to the best values of the process factors vary considerably from the simple heap situation to the far more expensive mechanized processes which are described later in Chapter 4.

TABLE 4 OPTIMUM COMPOSTING PARAMETERS

Parameter	Value
C/N ratio of feed	25 to 35/1
Particle size	10 mm for agitated systems and forced aeration 50 mm for long heaps and natural aeration
Moisture content	50 to 60% (higher values possible when using bulking agents)
Air flow	0.6 to 1.8 m^3 air/day/kg volatile solids during thermophilic stage, or maintain oxygen level at 10 to 18%
Temperature	55 to 60°C held for 3 days
Agitation	No agitation to periodic turning in simple systems Short bursts of vigorous agitation in mechanized systems
pH control	Normally none necessary
Heap size	Any length, 1.5 m high and 2.5 m wide for heaps using natural aeration With forced aeration, heap size depends on need to avoid overheating

CHAPTER 3

MATERIALS FOR COMPOSTING

3.1 GENERAL

Large quantities of organic matter need to be supplied to soils in the tropics and subtropics in order to provide plant nutrients, to help moisture retention and to keep the soil structure in good condition. Because the rate of organic matter oxidation is fast, due to high soil temperatures, frequent additions of compost are required, each year if possible. Hence it is most worthwhile to take care in saving organic waste so that it can be composted and recycled to help the soil in its task of food production.

Virtually anything which has once lived, of plant or animal origin, will decompose in a compost heap. Fresh green material breaks down very quickly, straws and woody material take longer; leather and bones are very slow and may need to be put through several heaps to reach complete breakdown.

In some communities many of the organic materials which might be composted are more important for other purposes. For instance, fresh green weeds and leaves are frequently used for animal fodder; straw from paddy (rice) and other cereals is used for the same purpose. In some parts of the world, wood and sun-dried dung are the only sources of fuel for cooking. In these latter cases, material for composting will become more readily available as tree-planting programmes are implemented to generate adequate firewood. Also, as the standard of living rises enough, capital should become available for the construction of biogas plants in which night soil and manure are converted to methane for fuel and biogas sludge for composting.

By contrast, in towns and especially in cities, the disposal of refuse and night soil or sewage sludge is becoming an increasing problem. The composting of these materials can supply organic matter for local farms and market gardens which grow the food for the urban population; the ease and cost of transport will limit the distance that the compost can be sent into the countryside.

Composting always proceeds better when a variety of organic wastes are being processed, rather than just a single material. Each component of the mixture makes its own contribution to the biological and nutrient cycles of the heap. For effective composting in which high temperatures are generated, the heap needs to be assembled in one operation if possible. For a smallholding in which not much organic waste is generated this may entail saving up wastes until sufficient accumulate, joining in a communal composting scheme or bringing in material from outside the premises.

This chapter describes the materials which are suitable for composting and their collection, storage and preparation prior to assembly into a compost heap.

3.2 MATERIALS

A compost heap needs a supply of mixed organic wastes, sometimes an activator to supply extra nitrogen and micro-organisms, a little soil and, if available, some recycled compost. The mass needs moistening, either with water or wet sludge/night soil and then exposure to air for composting to proceed.

3.2.1 Unwanted Materials

Some items should not be put into compost heaps. They take no useful part in the process, will not break down, and may cause problems for people handling the product in agriculture. Such materials include broken china, glass or pottery, pieces of metal or wire, plastic bottles and sheet plastic, rubber, nylon and other man-made fibres, fresh coal ash and soot from chimneys. Organic materials which should not be put into heaps are plants, twigs and fruit with hard thorns or prickles, and the roots of persistent perennial weeds which may not be killed by the heat evolved in the heap. Figure 12 illustrates some of these unwanted wastes.

3.2.2 Useful Organic Waste Materials

Large quantities of organic wastes are generated in nature. Thought must be given to collecting as much as possible from the following sources - home, garden, crops, livestock, forest, rivers, seas, industry, town and city refuse. Some of these materials are shown in Figure 13, and their nitrogen contents and carbon to nitrogen (C/N) ratios listed in Table 5.

3.2.3 Home

Domestic wastes include material from food preparation such as vegetable peelings and unwanted leaves, fruit skins, egg shells, tea leaves and coffee grounds. Scraps of fish and meat can lead to problems by attracting flies or vermin unless they are put into the hot centre of an active compost heap. A little paper and cardboard can be used if it has been torn up and soaked in water first. Rags of cotton or wool, and even pieces of leather, will eventually disintegrate. House dust, sweepings and ash from wood fires can all be used. Coal ash and chimney soot should only be included after they have been left in the open for 3-6 months.

A most important waste in the home is that from humans. It is vital that the collection and treatment are done hygienically to prevent the spread of disease and the causing of odours. In villages and towns which do not have waterborne sewage disposal systems, human waste is collected in the form of night soil. In many countries there is a prejudice against the use of night soil on crops; however, there is evidence that such prejudice is now weakening. In China night soil has been used for crop fertilization for centuries. The use of this waste is vital in closing the nutrient cycle (Figure 1), thereby preventing the loss of the major plant nutrients. This in turn reduces considerably the need for artificial fertilizers in a community where cash and transport are needed to bring them in, even if they are indeed available. In the United Kingdom, with its long established sewage system, about 40 percent of the sewage sludge produced is returned to agriculture.

3.2.4 Garden

Wastes from the garden include the remains of vegetable crops, dead flowers, stems, stalks, thin prunings and dry bonfire ash. Most garden weeds are of particular value; they are normally the only whole plants put into compost heaps and contain a wide variety of trace elements. However, the roots of the persistent, perennial weeds are best burnt. Bonfire ash should be covered until use; rain will rapidly leach out the useful nutrients, especially potassium.

A proportion of dry, fallen tree leaves, up to 20 percent, can be used but an excess should be avoided as they can greatly reduce the decomposition rate of other wastes. Where there is an excess of fallen leaves they should be gathered into a separate heap and allowed to decay over 1-2 years into leaf mould; this needs no attention apart from occasional watering.

Figure 12 Uncompostable wastes

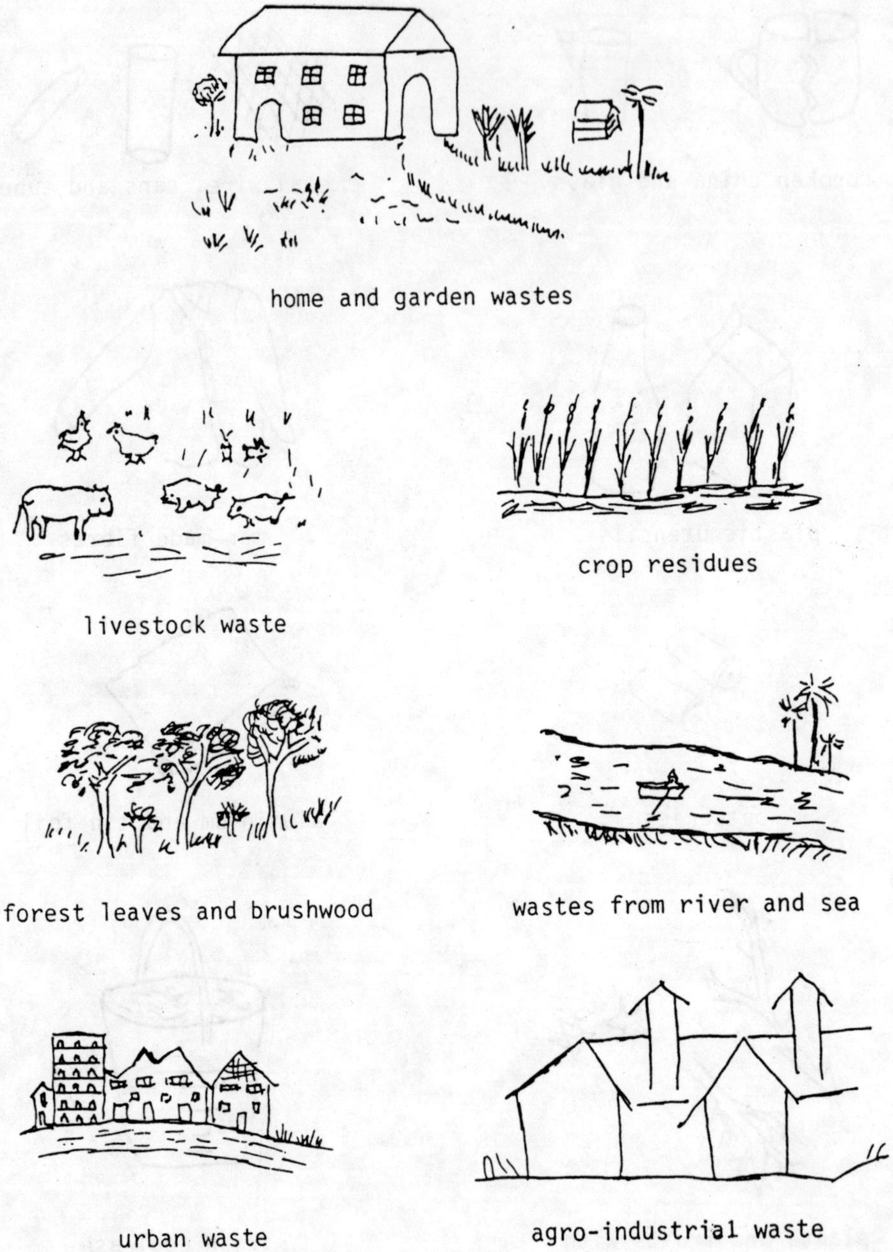

Figure 13 Sources of waste

Table 5 APPROXIMATE COMPOSITION OF MATERIALS SUITABLE FOR COMPOSTING

Materials	Nitrogen % dry weight basis	C/N ratio
Animal urine	15-18	0.8
Dried blood	10-14	3
Hoof and horn meal	12	ND
Fish meal	4-10	4-5
Oil seed cakes	3-9	3-15
Night soil	5.5-6.5	6-10
Sewage sludge, activated	5-6	6
Poultry manure	4	ND
Bone meal	2-4	8
Grass clippings, young	2-4	12
Green manure plants	3-5	10-15
Brewers waste	3-5	15
Urban refuse, high vegetable content	2-3	10-16
Coffee pulp	1.0-2.3	8
Water hyacinths	2.2-2.5	20
Pig manure	1.9	ND
Cattle manure	1.0-1.8	ND
Press mud cake	1.2-1.8	ND
Urban refuse, high paper content	0.6-1.3	30-80
Millet and pigeon pea stalks	0.7	70
Wheat straw	0.6	80
Leaves, freshly fallen	0.4-1.0	40-80
Sugar cane trash	0.3	150
Saw dust, fresh	0.1	500
Paper	nil	infinity

ND = not determined

The resulting mould is useful for mulching and for potting mixtures. Similarly, an excess of fresh lawn mowings is undesirable as they can settle quickly into a thick layer which prevents air movement through the compost heap. Such mowings should not exceed 50 percent of the heap and they must be mixed in with stalky, strawy material to keep the mass well opened out.

As much soil as possible should be shaken from the roots of garden plants before they are put into the heap as too much soil will slow down the composting process.

To provide extra material for the compost heap annual plants such as sunflowers (Helianthus annuus L.), and perennial plants such as Russian comfrey (Symphytum X uplandicum Nym.) and sunn hemp (Crotalaria juncea) can be grown. Such plants should be used just before they form seeds. 1/

As mentioned earlier, material with hard prickles or thorns should be kept out of the heap.

3.2.5 Crop Wastes

A wide range of crop wastes from farms, market gardens and plantations are excellent for composting if suitably mixed; they are not so good if employed singly.

The wastes from cereal crops such as paddy (rice), wheat, barley, millet and sorghum are useful, not only the straws but also the husks, chaff and bran obtained on threshing and milling. Maize (sweet corn) straw is especially helpful as it grows tall and provides a lot of material. The leaves and stems of the pulses such as chick peas, pigeon peas and mung beans, and oil seeds plants such as castor, groundnut, linseed and sunflowers all supply suitable wastes. The stalks of cotton, jute and tobacco and sugarcane trash may all be used. The leaves of shrubs and trees such as arecanut, coconut and banana can also be collected (Plates 1, 2 and 3).

Most of these materials are from mature plants or are fallen leaves. Consequently their C/N ratios are high and they need to be mixed with wastes having low C/N values, such as manures, in order to compost properly.

3.2.6 Livestock Wastes

The manures, dung, urine and droppings from all types of animals and poultry are excellent for composting (Plate 4). Not all the major, minor and trace nutrients in feed and fodder are used by the animal for body maintenance and the production of meat, milk, wool and eggs; some are excreted in the form of body wastes and will normally be a large percentage of the total nutrients consumed. A similar relationship exists between humans, their food, and night soil. As shown in Table 5, manures provide material of low C/N value for blending with the crop wastes mentioned in the previous section. If such manures were composted by themselves, air would fail to penetrate the mass which would quickly go anaerobic and cause bad smells; in addition much nitrogen, a valuable plant nutrient, would be lost in the form of ammonia gas.

Hence, where possible, crop wastes such as straws should be used in animal sheds as bedding. In this way they will soak up urine, a rich source of potassium, and dung. Deep littering is more effective in providing material for recycling than is daily cleaning of dung from bare floors. Where sufficient bedding is used for cattle, pigs and poultry in the tropics, there appear to be no ill-effects on animal cleanliness and health, and it does not create a fly breeding problem.

Sometimes there is only sufficient straw for feeding the cattle and none left over for bedding. In such cases, earth can sometimes be removed from silage pits and threshing floors, and silt recovered from water supply tanks and channels; this is stored close to the cattle shed. This earth or silt is then spread on the floor of the cattle shed to absorb the urine and is renewed every three or four months; by this time it will have been trodden into a compact mass by the cattle and will need breaking up. If they are available, dry sawdust and wood shavings can be used in a similar manner.

1/ Some plant names are given in Latin in Appendix 8

Plate 1 Husks, stems and dried leaves should be gathered up and used for compost
Source: FAO filmstrip on Compost - Africa

Plate 2 Leaves from trees, shrubs and plants make useful material for the compost heap
Source: FAO filmstrip on Compost - Africa

Plate 3 Green wastes such as freshly cut grass, leaves and weeds are gathered ready for composting
Source: FAO filmstrip on Compost - Thailand

Plate 4 Manures in compost heaps lower the carbon to nitrogen ratio and speed up the breakdown of straws and stalks
Source: FAO filmstrip on Compost - Africa

The object of such cattle shed management is to recover the nutrients for crop production instead of letting them soak away uselessly into the ground below the shed. Not only are the manures recycled but also spoilt or wasted animal fodder.

3.2.7 Forests

Large amounts of organic material should normally be available from forests and wooded areas. Unfortunately, in many tropical countries population pressure has created a greater demand for firewood than local trees can supply. In consequence, extensive deforestation has taken place, often resulting in hillside soil erosion. This situation can only be improved by tree planting programmes.

The other problem has been that of shifting cultivation using the slash-and-burn mechod. In this, areas of forest are cleared for cultivation but all the trees, shrubs and plants are burnt. As a result, there is a total loss of the organic matter; although the phosphorus, potassium, minor and trace nutrients are mainly saved in the ash, the major nutrient, nitrogen, is almost entirely lost. This technique represents a tremendous loss of resources. The plot can often only be cultivated for a few years, then a new site has to be chosen, leaving the impoverished one to lie fallow and slowly reclothe itself with vegetation. This practice may have been tolerable in the past but population pressure now forces farmers to shorten the fallow period and thus insufficient time is available for natural regeneration. The result is a rapid and serious breakdown in soil fertility.

Unfortunately, it is exactly these hard mature woody wastes from forest trees and shrubs, containing much cellulose and lignin, which are so important in composting in the tropics. Fresh green crop wastes break down very quickly in the compost heap; these provide the sugars and simple organic compounds which lead to the great increase in the numbers of micro-organisms and the evolution of heat which causes the temperature to rise sufficiently for disease organisms to be killed. However, the resulting compost degrades or mineralizes quickly in the soil, releasing the required plant nutrients; it probably has little long-term effect on soil organic matter in the tropics. Such green wastes need to be complemented by woody wastes which only decompose a little in the compost heap but which continue to break down in the soil over many years, providing a reservoir of organic matter for enhanced water retention and improved soil structure. Recent work on the composting of brushwood and of wood chips from larger tree branches shows that these materials will break down if care is taken.

Consequently, when more trees are planted and the firewood supply assured in a local area, forest thinnings and litter can become an important source of organic waste material for composting.

3.2.8 Rivers and the Sea

A number of useful materials for composting can be obtained from rivers, canals, ponds and the sea. The most important is naturally water because compost heaps need a moisture content of 50-60 percent, either as water bound up in plant tissues or present in manures and night soil, or as water added during formation of the heap and any subsequent turning.

Green aquatic plants are often available in quite large quantities. The most notorious is the water hyacinth (Eicchornia crassipes) which spreads extremely rapidly, transpires enormous quantities of water which could otherwise be used for crops, and can block waterways to passage of boats. Being of very high moisture content such plants need to be wilted for several days before composting if they are being used in large quantities.

In areas close to the sea, supplies of seaweed of several types are often an important source of organic material. If handled skilfully seaweed can form up to about 50 percent of the material in a compost heap. It needs first to be drained of seawater. If the compost is to be used on heavy clay land the seaweed should ideally be washed with fresh water to remove its salt content; for other soils this is rarely necessary. The seaweed is then well mixed with strawy crop wastes to provide absorbent material and an open matrix; the mixture is next spread 300 mm deep for 2 days during which the weed loses about half its moisture content by drainage and evaporation. It is then blended with further vegetable and crop wastes and built quickly into a compost heap. If seaweed is mishandled it can degrade rapidly into a slimy, smelly, anaerobic mass which quickly attracts flies.

Silt can be recovered from waterways and water tanks; it can be used in the cattle shed or directly in, or over the top of, the compost heap. Some composting techniques in China use large quantities of silt when composting night soil, water hyacinths or grass. Silt obtained from glacial rivers is especially rich in nutrients; the highly fertile terraced fields of the Hunza Valley in the Himalayas were built from such silt taken back up the mountain side.

In ponds which are used for intensive fish culture large amounts of plant, animal and microbial debris sink to the bottom; the pond mud thereby becomes rich in both organic matter and in the plant nutrients. In China, such fish ponds are located between cropped areas; the mud is collected 4-5 times a year and used in crop production.

3.2.9 Urban Wastes

The countryside produces most of the food for the inhabitants of towns and cities; in turn the latter generate roughly comparable quantities of urban wastes in the form of refuse and night soil or sewage sludge. To complete the cycle of nature such wastes should be returned to the countryside for mixing into the soil.

In many parts of the world the population is still increasing rapidly; not only are there intermittent shortages of food, there is also extensive rural unemployment and a drift of people from the countryside into towns and cities. As such centres of population grow so do the problems of urban waste collection and transport over lengthening distances to the countryside. The future will definitely see increasing quantities of urban waste being generated and greater difficulties in its disposal.

Urban refuse comprises the solid wastes from human dwellings; it contains food wastes, paper and cardboard, cinders and ash from heating systems, containers in glass, plastic or metal, plus derelict furniture and household appliances. In addition there are substantial quantities of commercial refuse from offices, shops and markets together with industrial and construction waste.

The composition of urban refuse from cities in different regions of the world is given in Table 6; considerable local variations are to be expected. The figures indicate that in the new cities of the developing world there is a greater contribution from vegetable and putrescible wastes while in Europe there is a greater output of paper and of containers of all sorts. In poorer countries there is undoubtedly a large amount of salvage and recycling of metals and other materials which have any value.

Table 6 TYPICAL COMPOSITION OF URBAN REFUSE FROM DIFFERENT PARTS OF THE WORLD

Refuse fraction	% by weight (fresh weight basis)			
	Middle East	Far East	South America	Europe
Vegetable & putrescible material	50	75	55	16
Paper	20	2	15	43
Plastics & textiles	10	4	10	7
Glass	2	0	4	10
Metals	10	0	6	10
Miscellaneous	0	7	10	1
Inerts & rubble	8	12	0	13
Approximate moisture content within the total	42	60	45	18

In refuse only the vegetable/putrescible and paper fractions are of any value in composting; all other materials are unwanted and can cause problems to the composting process or to those using the compost product. Hence Table 6 shows that the refuse from Europe is much less suited to composting and needs more separation of unwanted components.

Grubben (1982) describes the practice of certain market gardeners close to two cities in Benin, West Africa, who incorporate large quantities of fresh or only slightly decomposed refuse into their soils prior to growing leafy vegetables. An extension project to teach compost making proved a complete failure, apparently due to the high labour input required to make mature compost. Nevertheless, use of raw refuse cannot be recommended for general practice because it is unhygienic; even composting city refuse for a few days, to bring it to its temperature peak, will greatly reduce the number of disease organisms (pathogens) which are present.

In addition to refuse, the other component of urban wastes important to composting is night soil, or sewage sludge where there is a waterborne sanitation system. There is an urgent need to recycle this material, with its wealth of plant nutrients, back to agriculture if the need for mineral fertilizers is to be reduced. However, night soil and sewage sludge contain many of the pathogens to which the human body is prey and hence care must be taken to prevent such disease organisms affecting workers and food crops. Partial control can be achieved in biogas digesters and almost total destruction by composting. The sludge from such biogas digesters is a valuable addition to the compost heap (Plate 5).

In many of the less developed countries where cities have grown rapidly and there are insufficient funds for adequate waste disposal, the material collected is often dumped in pits, quarries and hollows close to the city. This not only represents a loss of organic matter and plant nutrients to agriculture but can cause pollution of water supplies and give rise to problems of smell, flies, vermin and scavenging animals and birds. Ultimately, with sufficient finance and organization these materials will need to be separated and the compostable fraction returned to the countryside for soil enrichment.

Plate 5 Biogas sludge removed periodically from a digester is a good material to add to a compost heap
Source: FAO filmstrip on Uses of Coffee Pulp

Plate 6 Coffee pulp is an important waste in the coffee industry; it can be composted by itself or, preferably, with other wastes
Source: FAO filmstrip on Uses of Coffee Pulp

3.2.10 Agro-industrial Wastes

The agro-industries process the primary products from smallholdings, farms, plantations, forests and animal slaughter houses into foodstuffs and products, either for return to the grower or for distribution into local, national or international markets. The by-products and wastes arising from these operations are available in substantial quantities, often as large as the desired foodstuff or main product. Some of these materials have already been mentioned under Section 3.2.5 Wastes, where they are left on the smallholding or farm and not transported to a processing factory. Wastes from such factories are often poorly used or are dumped, leading to environmental pollution problems and a loss of plant nutrients and organic matter which should otherwise be returned to the soils growing the primary products.

As shown in Table 7 agro-industrial wastes come from the growing and processing of a wide range of fruits, nuts and vegetables, beverages such as cocoa, coffee and tea; cereals such as paddy, maize, wheat, millet and sorghum; pulses such as beans and peas; fibres of cotton, wool, jute and silk; sugar; meat, fish and marine products; nuts and seeds for vegetable oils; tobacco and saw mill wastes.

Table 7 AGRO-INDUSTRIAL WASTES

Vegetables	Decayed cabbage and lettuce leaves; tops of root vegetables; wastes from tomato canning
Fruits	Trimmings from pineapples; banana stems; shells from cashew nuts, groundnut, wood apple; peel and stone of mango; peel and pips from citrus and passion fruits
Cereals	Husks and bran from paddy, wheat, sorghum (jowar) and barley; maize cobs
Beverages	Brewery and wine wastes; cocoa pods; coffee pulp; tea stalks and sweepings
Fibres	Cotton gin wastes; coconut husk; coir dust; jute mill wastes; silkworm wastes
Sugar	Bagasse; press mud cake; vinasse (the liquor from industrial alcohol distillation)
Vegetable oil cakes	Oil cakes from copra (coconut); castor; seeds of cotton, groundnut; karanj; linseed; mahua; mustard; neem; oil palm kernels; rape; sesame; sunflower 1/
Tobacco	Tobacco seed cake, leaf scrap and stalk
Meat	Bones, hair, horn, hooves, feathers, blood, leather and fat
Fish, river and marine products	Fish wastes; surplus and inedible fish; prawn wastes (shell and head); frogs and seaweeds
Timber saw mills	Sawdust; wood shavings; wood chips
Factory effluents	Washings and sludges from various industrial processes

1/ Karanj = *Pongamia glabra*; mahua = *Madhuca longifolia* (L.) Macbr.; neem = *Azadirachta indica*.

Some of these wastes have significant value as animal feedstuffs either locally or for export to animal feed compounders in the West. Animal products are often processed into fertilizers of reasonable nitrogen and/or phosphorus content. In sugar production most bagasse is normally used as fuel in the factory boilers. Other wastes are sometimes employed as raw materials for industrial processes; an example is the use of cocoa pods in Ghana as a source of potash for the manufacture of soap. The remaining wastes, especially in small local agro-industries, should be available for compost production either centrally at the factory or distributed to nearby smallholders for incorporation into their heaps. Plate 6 shows coffee pulp which is an example of such wastes.

3.3 CHEMICAL COMPOSITIONS

The chemical compositions of organic wastes vary enormously with different contents of moisture, carbon, hydrogen and oxygen, the major plant nutrients nitrogen, phosphorus and potassium, the minor nutrients and the trace nutrients. These nutrients collectively form the mineral matter. Even the composition of a particular plant waste will vary slightly, depending on the plant variety, the soil on which it is grown and the agricultural practices employed, such as irrigation and use of fertilizers.

Table 5 listed the approximate composition of a number of organic waste materials. It gave the percent nitrogen on a dry weight basis of the material and the carbon to nitrogen, C/N ratio. The moisture content of the waste is not very important; it varies enormously depending on exposure to sun, wind and rain and it can be adjusted to the desired level in compost heap construction by the addition of water. The nitrogen content is fairly important, being the major plant nutrient. However, the C/N ratio should be considered in planning composting operations. For efficient compost production the mixed wastes should have a C/N ratio in the range 25-35/1 as stated in Sections 2.4.4 and 2.4.11. Accordingly, where a particular waste has a ratio lower than the desired value, then it should be mixed with material having a ratio higher than 25-35/1. For example, animal manures or night soil with low values need to be blended with plant straws and stalks with high ratios.

3.4 QUANTITIES

The amount of plant material which can be generated in Nature under ideal growing conditions is enormous. On deep fertile tropical soils with adequate nutrients and irrigation at least 150 t/yr/ha could be produced from appropriate plants and trees. In a tropical rain forest the material eventually falls to the ground and decays, thereby achieving a closed cycle of nutrients and organic matter. If such quantities were removed from a particular area, soil fertility would rapidly decline unless sufficient organic matter and plant nutrients were returned to the soil to compensate for those removed. If this were not done, productivity would decline, often very rapidly on some tropical soils. Eventually, productivity would become almost nil, as is now the case in parts of Africa where the top soil has been washed from hillsides or where desertification has taken over and crops have virtually failed; food aid then becomes essential for survival and practically the only organic wastes available are those from humans.

It is impossible to quantify exactly the organic wastes generated per head as many factors, including geography and affluence, are involved. As an indication, however, Table 8 taken from Gaur (1978) has been included; this gives an idea of the organic wastes generated by the agro-industries of India, a country with a population then of about 600 million people.

Table 8 SOME AGRO-INDUSTRIAL ORGANIC WASTES IN INDIA

Waste		Quantity tonnes/year
Vegetables and fruit:	1. Mango peel and stone	25 000
	2. Citrus peel	400 000
Cereals:	1. Paddy husk	15 000 000
	2. Paddy bran	2 500 000
Beverages:	1. Tea waste	10 000
	2. Coffee husk	45 000
Fibres:	1. Cotton stalks	12 000 000
	2. Cotton dust	30 000
	3. Jute sticks	2 500 000
Sugar:	1. Bagasse	5 300 000
	2. Press mud cake	200 000
Vegetable oils:	1. Edible oil cakes	2 000 000
	2. Neem seed	1 000 000
	3. Mahuja seed	70 000
	4. Karanj seed	26 000
	5. Sal seed	25 000
Tobacco:	1. Leaf scrap and stalk	34 000
Meat, fish & marine products:	1. Blood	55 000
	2. Bones	450 000
	3. Prawn waste	40 000
Sawmills:	1. Sawdust	2 000 000

Source: Gaur (1978)

Table 9 POTENTIAL SUPPLY OF ANIMAL, RURAL AND CITY WASTES FOR RECYCLING AS FERTILIZERS IN INDIA

Waste	Potential supply million tonnes/year	Nutrients in million tonnes/year (% nutrient content on fresh weight basis)		
		N	P	K
Animal dung (wet)	960	1.91 (0.20)	0.42 (0.04)	1.20 (0.13)
Animal urine	370	2.22 (0.60)	0.02 (0.005)	1.54 (0.42)
Rural compost	600	3.00 (0.50)	1.32 (0.22)	2.49 (0.42)
Urban refuse compost	15	0.23 (1.50)	0.07 (0.47)	0.19 (1.27)
Sewage for 36 million people of urban population	1 460 million m^3 per year	36 500 tonnes/year	3 200 tonnes/year	18 200 tonnes/year

Source: Gaur (1978)

The quantities of urban waste generated will depend on many factors, including the standard of living of the people. For refuse the average for 16 Indian cities listed by Gaur (1978) is about 0.18 tonnes per year per person. In Shanghai, in China, the amount is about 0.24 tonnes per year per person. By contrast the average over the UK is 0.33 tonnes per year per person, with roughly an equal amount of commercial refuse in addition.

The amounts of human wastes which can be recycled depend on whether they are collected as night soil or in a waterborne sewage system. In the UK, which has a waterborne system covering most of the population, the amount of sewage solids is about 0.02 tonnes per year per person.

In Table 9 are data from Gaur (1978) summing up the potential of animal, rural and city wastes for recycling as fertilizers in India with its population then of about 600 million people. In Table 10 similar data are given for China from FAO (1978a); in a closely organized society about 70 percent of these wastes are actually recycled. In this table the category of 'other materials' includes items such as ashes, fish wastes and silk wastes. China has a population of about 900 million of which over 80 percent live in rural areas. Table 10 is based on the data given in Table 11 for material generated per head, per animal or per hectare.

Table 10 POTENTIAL SUPPLY OF ANIMAL, RURAL AND CITY WASTES FOR RECYCLING AS FERTILIZERS IN CHINA

Waste	Potential supply Million tonnes/year	Nutrients in million tonnes/year (% nutrient content on fresh weight basis) N	P	K
Cattle manure	598	3.58 (0.60)	0.78 (0.13)	3.95 (0.66)
Pig manure	727	3.63 (0.50)	1.28 (0.18)	3.02 (0.42)
Goat and sheep	110	0.66 (0.60)	0.15 (0.13)	1.09 (0.99)
Poultry manure	34	0.49 (1.46)	0.17 (0.51)	0.17 (0.51)
Plant residues and water plants	397	1.19 (0.30)	0.35 (0.09)	1.98 (0.50)
Green manures	61	0.24 (0.40)	0.03 (0.04)	0.20 (0.33)
Oil seed cakes	5	0.04 (7.00)	0.003 (0.44)	0.01 (1.66)
City refuse	22	0.13 (0.60)	0.04 (0.18)	0.18 (0.83)
Night soil	216	1.29 (0.60)	0.19 (0.09)	0.54 (0.25)
Mud and silt	183	0.46 (0.25)	0.11 (0.06)	0.23 (0.12)
Others	10	0.04 (0.40)	0.01 (0.09)	0.03 (3.32)
Total	2 363	11.75	3.11	11.40

Source: FAO (1978c)

Table 11 QUANTITIES OF WASTES GENERATED PER INDIVIDUAL UNIT IN CHINA

Waste	Quantity
Cattle manure	6 tonnes per year per animal
Pig manure	3 tonnes per year per animal
Goat and sheep manure	0.8 tonnes per year per animal
Poultry manure	1.0 tonnes per year per 40 birds
Plant residues and water plants	3.0 tonnes per year per hectare
Green manures	10.0 tonnes per year per hectare
City refuse	0.15 tonnes per year per person
Night soil	0.25 tonnes per year per person
Mud and silt	10 million hectares in delta regions receive 8 t/yr/ha and 50 million hectares in other areas receive 2 t/yr/ha

Source: FAO (1978c)

China is probably the country which most effectively recycles its organic wastes back to agriculture. If the area of cropped land is assumed to be 130 million hectares then the average quantity of organic fertilizers employed, assuming some 70 percent is actually recycled, is about 13 t/yr/ha. In terms of the major plant nutrients each hectare of cropped land annually receives 64 kg of N, 17 kg of P and 61 kg of K through organic fertilizers. However, the usage of these manures is already approaching the maximum quantities which can be recycled economically; hence increased employment of mineral fertilizers has been necessary to improve crop production.

3.5 COLLECTION, STORAGE AND PREPARATION

As previously described, organic wastes suitable for composting cover a wide range of materials from highly putrescible animal manures, human wastes and rotting fruit, to tough woody stems which are fairly resistant to decay. Some wastes are generated infrequently; examples are the straws from cereals which only arise at harvest time and fresh fruits which are available only when in season. Other m terials are generated continuously; examples are manures from cattle sheds, night soil and city refuse.

Compost heaps can be assembled infrequently or continuously, depending on the quantity of wastes and the labour available; farm compost heaps can best be built during periods when labour is not required for other agricultural operations.

Accordingly, thought must be given to collecting the wastes and transporting them to the composting site in an efficient manner. Where compost heap construction is intermittent, care must be exercised in storage of the wastes. It is important that most of the breakdown takes place in the compost heap, thereby releasing much heat and generating high temperatures; premature breakdown during storage is not helpful to effective composting. In addition, the storage of putrescible wastes should be done with care, to avoid causing bad smells and giving opportunities for fly breeding or attracting vermin.

Some materials are not in a suitable state for composting and require a degree of preparation before being built into heaps.

3.5.1 Collection

Materials intended for composting need to be collected frequently from their place of origin, whether this be the home, field, city or industry and transported to the composting site for heap construction or temporary storage. This is both to reduce environmental pollution by smell, flies and vermin and to reduce breakdown and leaching of plant nutrients.

Equipment for collecting material for small-scale and farm-scale heaps is discussed in Section 4.6.

For the collection of urban refuse from towns and cities, vehicles vary from hand carts, tricycles with cane baskets or metal bins, tractors and trailers, to sophisticated modern vehicles with facilities for refuse compaction. In China city refuse is often transported daily to the countryside in barges that move along the network of canals; it is unloaded at the water's edge on commune fields where it is composted.

Night soil from private and public latrines in India is usually collected in closed buckets mounted on wheelbarrows (Figure 14); in a city it is taken to local semi-underground tanks. The tank contents can then be removed by a vacuum tanker vehicle and transported to the disposal site. In China night soil is conveyed to rural areas by closed carts or by boats.

Silt from the bottom of waterways and lakes in China is removed by vacuum pumps mounted on boats and deposited in pits on the banks prior to use in compost heaps.

Figure 14
Containers for night soil collection

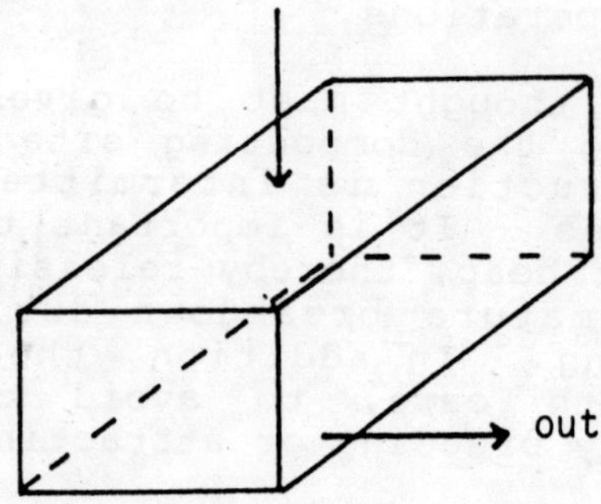

3.5.2 Storage

Materials for composting frequently need to be stored in order to fit in with the timing of compost heap preparation. Some decomposition during storage cannot be avoided, particularly with highly putrescible materials; however, such decomposition should normally be kept to a minimum so that most can take place in the actual compost heap. Keeping material cool and dry and preventing ready access of air will reduce premature breakdown.

Putrescible materials such as fruit and vegetable wastes can be stored in a pit, or in a container above ground during the rainy season. They must be kept covered to prevent fly breeding and keep out rain. Nevertheless they will exude moisture into the soil; as this will contain useful plant nutrients the soil can be taken up when the pit is emptied and replaced with a fresh layer.

Green materials such as weeds and green manures, and especially aquatic weeds like water hyacinths, are best exposed to the sun and withered for at least two days to reduce their moisture content. They are then stacked, preferably under cover.

Very hard and woody material such as bagasse, maize and millet stalks, wood shavings, saw dust, and also waste paper, should be stored in an empty compost bin or pit with a little soil and kept moist; this treatment helps moisture soak into the material, making it easier to compost.

Night soil is the material which needs much care in storage to prevent pollution problems arising and to reduce pathogen risks. In China two methods are commonly practised. The material is stored in a properly designed closed storage chamber for a period of four weeks prior to use; this gives an opportunity for many pathogens and intestinal worm eggs to die, though this is not completely effective. Alternatively, if more money is available, the night soil is put into a proper anaerobic digester or biogas plant with a retention time of several weeks; not only is biogas generated but many pathogens and worm eggs are killed.

Where cities have waterborne sewage systems the sludge resulting from the treatment process can be kept in tanks and used after drying on sludge drying beds.

A material of major importance in composting is water because the desired moisture content of the assembled heap is about 55 percent on a fresh weight basis. This means that if the organic wastes are very dry then roughly an equivalent weight of water is needed. The moister the wastes, the less water is needed. Accordingly, composting on a large scale needs to be done close to an adequate supply of water, either alongside waterways, near wells or a water storage tank. Care must be taken to ensure that the composting material does not pollute drinking water sources.

3.5.3 Preparation

Some organic wastes are not in a suitable state for composting when in their normal form. Particle size is an important process factor as the micro-organisms need plenty of surface area for their attack; a size below 50 mm is desirable. Hence the branches of trees and shrubs, and the stalks, stems and straws of cereals, need breaking up into smaller lengths. Very rarely is power available for a mechanical cutter. One alternative is to put the wastes out on a farm road and let wheeled traffic crush it (Plate 7). Weak stalks and straws can be chopped with sickles or a hand-operated chaff cutter; hard woody material will need chopping with a machete or panga. Plate 8 shows compacted urine-earth from cattle sheds being crushed before use.

Plate 7 Crushing stems by road traffic before composting
Source: Howard and Wad (1931)

Plate 8 Crushing urine-earth f᷊ n cattle sheds in a lime mortar mill
Source: Howard and Wad (1931)

Hard dry woody materials also need to be well moistened before assembly into a compost heap; waste paper needs similar treatment. They can be stored moist in a pit as mentioned in Section 3.5.2 or soaked in water for about two days immediately prior to use.

Urban refuse needs some degree of preparation in order to remove uncompostable materials such as glass, metal and plastic. This process is important, not only to recycle such materials where possible, but also to prevent agricultural soils from slowly accumulating unwanted undegradable debris. The removal can be done by hand from the initial wastes where labour willing to undertake the task can be recruited. Alternatively, the refuse can be composted in its original form and unwanted material screened out of the product. In cities with fully mechanized composting plants such uncompostable items can be removed using a number of separation devices.

3.6 ADDITIVES

When a mass of mixed organic waste from several sources, containing some fresh green material, and of suitable particle size range and moisture content, is composted in accordance with the optimum process parameters listed in Table 4, composting is usually rapid and a satisfactory product results. The further one strays from the ideal conditions of raw material and processing, the slower the composting process and the worse the product. In these cases the addition of other materials is often justified in order to improve chemical composition and physical structure of the heap, to supply desirable micro-organisms and to reduce nitrogen losses. Moreover, it is sometimes desirable to enrich the compost product with extra plant nutrients.

3.6.1 Activators

The optimum C/N ratio of a composting mass is 25-35/1 at the start. It should contain a reasonable mixture of material which is rapidly broken down, some which is more slowly attacked in the heap and some which is woody, lignified and fairly resistant to attack in the heap but breaks down slowly in the soil afterwards.

If too much strawy, stalky and woody material is present, the C/N ratio is likely to be much greater than the optimum value. The rate of reaction will then be slow, adequate temperatures for pathogen kill probably will not be reached and a lot of organic material will be oxidized by the micro-organisms in an attempt to reduce the C/N ratio towards 10/1. This will much reduce the amount of product produced; even then the product is likely to rob nitrogen from the soil when added. To prevent this occurrence, it is important to reduce the C/N ratio of the initial material by the addition of an activator containing extra nitrogen in a fairly reactive form. Materials with low C/N ratios at the top of Table 5 are needed such as urine, dried blood, night soil, sewage sludge, manures and young green plant growth. Although hoof and horn meal has a high nitrogen content it is not suitable as an activator as the nitrogen is only released very slowly.

If organic materials are not available for use as activators nitrogenous mineral fertilizers can be used, such as urea, ammonium nitrate, ammonium sulphate and calcium cyanamid. Experience in the past appears to indicate that these are not so effective as organic sources of nitrogen; the latter apparently coat the surface of the organic waste with a film of colloidal particles which help the retention of moisture on the surface, thereby aiding the micro-organisms in their attack.

Even sawdust and wood shavings, which have little nitrogen present and contain mainly resistant cellulose and lignin, will break down in a few months if a source of nitrogen and of phosphate is added.

Other materials which will activate or speed up the composting process, particularly during the initial few days when heat generation is important, are ones supplying sugars. These will cause a rapid increase in the population of the micro-organisms. Two products from the sugar processing industry have value in this respect. One is press mud cake obtained from molasses filtration; the other is vinasse, the effluent liquor from molasses conversion to industrial alcohol.

3.6.2 Inoculants

As composting results from the activities of micro-organisms, it might be expected that the process would be improved by the addition of inoculants of special bacterial cultures. At one time this idea was derided and some experimental proof obtained on a laboratory scale to show that no advantage was obtained. Micro-organisms develop extremely rapidly and within a few days their numbers can build up to the maximum allowed by the environmental conditions in the compost heap.

Recent work, however, indicates that the recycling of a small quantity, 1-2 percent by weight, of product compost from a previous heap will supply a population of acclimatized micro-organisms to the fresh wastes being assembled into a new heap. This beneficial effect with acclimatized micro-organisms has been demonstrated when composting municipal wastes and with plant materials such as evergreen leaves of holly, ivy and pine needles, which are normally difficult to degrade.

Work in India has also shown that inoculating a compost heap with a suitable strain of nitrogen-fixing bacteria Azotobacter in the presence of added rock phosphate will significantly increase the nitrogen content of the final compost. The inoculation is done once the heap has passed through the thermophilic stage of composting and is cooling down.

3.6.3 Other Materials

The addition of finely ground rock phosphate or calcium phosphate, up to 2 percent by weight, even without Azotobacter, has a beneficial effect both in speeding up the composting process and in nitrogen conservation. The first effect is believed to be due to stimulation of the micro-organisms which attack the cellulose. The second effect is due to stimulation of the nitrogen-fixing bacteria Azotobacter which are present and a suppression of the denitrifying bacteria which break down the nitrate form of nitrogen.

The presence of 1-2 percent by weight of soil, particularly clay in a dry powdered form, is beneficial when sprinkled on the layers of wastes. It not only helps by introducing soil micro-organisms into the heap but also by holding on to ammoniacal nitrogen during the thermophilic stage until it can be used during cellulose decomposition later. The Chinese go much further and use larger quantities of river mud and silt, both within their heaps and plastered over the outside where it helps with heat insulation and odour control.

When the organic waste consists entirely of finely divided solids such as partially dewatered sewage sludge, it is virtually impossible for air to penetrate the mass. In such cases a bulking agent such as wood chips or fragments of rubber tyres is necessary to open out the heap structure. The bulking agent is sieved out of the compost product and recycled, though with some loss by degradation in the case of wood chips.

Many writers of books on composting have laid emphasis on the need to incorporate in a compost heap neutralizers or chemical bases such as calcium carbonate or lime. They postulate that because the initial products of composting are the simple organic acids and the mass goes slightly more acid to a pH of about 5.5, then chemical bases are needed to raise the pH to the level of 7.0 to 8.0 which favours the growth of bacteria and fungi. However, as shown in Figure 11 the mass normally starts to turn alkaline at the end of the mesophilic stage when ammonia is formed on the breakdown of protein material. The presence of chemical bases at this stage would help to liberate this ammonia from the heap. Hence the present authors can see no justification for adding chemical bases to normal compost heaps.

3.6.4 Compost Product Enrichment

Compost is essentially a soil conditioner containing a fairly low level of plant nutrients. Where food production needs to be increased and NPK mineral fertilizers can be afforded and transported to the locality, then there is some incentive for passing these through the compost heap. Care must obviously be taken with nitrogenous fertilizers to ensure that excess is not added, leading to nitrogen loss as ammonia. Nevertheless, the addition of nitrogen, particularly to mixtures with high C/N ratios, will reduce the amount of organic matter oxidized to reach the final C/N ratio, thereby increasing the amount of product compost.

As mentioned above, the addition of phosphate has a beneficial effect on a compost heap. The composting process makes the phosphate more water-soluble and hence more available to plants. Cheap rock phosphate can be used instead of the more expensive superphosphate which has been chemically treated to make it more water-soluble.

The addition of extra potassium fertilizer has no apparent beneficial effect within a compost heap. However, cheap forms of potassium rock can be added and will be made rather more water-soluble and more available to plants from the product compost.

The compost heap can be used to rectify any major deficiencies in trace nutrients in local soils. An example is that of the sub-Himalayan belt stretching through Bhutan, Nepal and North India where soils are often highly depleted in iodine, with a consequent bad effect on the health of the local inhabitants. Recent investigations have indicated that in certain villages where human wastes are used for fertilizing crops, a single course of iodine injections in the population some 15 years earlier had a notable effect on village health in ensuing years. In such localities the addition of potassium iodide (KI), or other iodine-bearing material, to the compost heap in small amounts would doubtless have a similar beneficial effect. Deficiencies in other essential trace nutrients could be corrected in a similar manner.

CHAPTER 4

PRACTICAL COMPOSTING PROCESSES

4.1 GENERAL

This chapter describes practical methods of composting the organic wastes which are discussed in Chapter 3. The techniques try to employ as closely as possible the optimum process conditions set out in Section 2.4.11. The key to successful composting is that the process adopted must be simple and workable with the labour, the skills and the finances available. In some instances in the past attempts to teach composting have failed because too perfect a product has been sought, involving many turnings of the heap; this has taken up too much labour at a time in the farming year when it was urgently needed for other agricultural operations.

In deciding on the process to be employed one must consider:

i. the type of wastes available, their ease of breakdown, and whether they are likely to contain many pathogens;

ii. the quantity of material requiring processing;

iii. the allowable cost in terms of labour, equipment and space;

iv. the use to which the compost will be put.

Materials which are highly putrescible break down quickly, generate much heat and reach high temperatures with fairly small quantities of wastes. By contrast, stalky and woody materials release their heat more slowly and hence require larger quantities in order to produce adequate temperatures for pathogen destruction.

With the quantity of material available one normally needs to build as large a heap as possible provided that air can penetrate into the mass. The maximum size should normally be not more than 1.5 m high and 2.4 m wide in cross-section; the height will decrease considerably as composting proceeds.

The outer edges of a composting mass will be cool, ranging from ambient air temperature on the surface to the core temperature about 150 mm inside. Hence the larger the mass, the greater the volume of the central core which will achieve adequate temperatures for pathogen and weed seed destruction. This effect is shown in both Table 12 and Figure 15. Small heaps have a larger surface area to volume ratio than large heaps and thus lose relatively more heat. Even with a cube-shaped heap having sides 3 m in length the effect of the outer cool layer is very marked; with a heap having sides only 0.3 m in length there is no central hot core at all. This shows that with small heaps much benefit can be obtained from side wall protection to improve insulation and thereby decrease heat loss; it will also prevent cooling and drying by wind blowing on the sides of the heap.

The quantity of material to be processed will also govern the size of the heap; wastes from the home are generally only sufficient for a small heap whereas the greater amounts of urban waste from a town or city will need several large heaps. Wastes containing night soil also require a big heap because the large central core should reach temperatures capable of achieving an adequate pathogen kill.

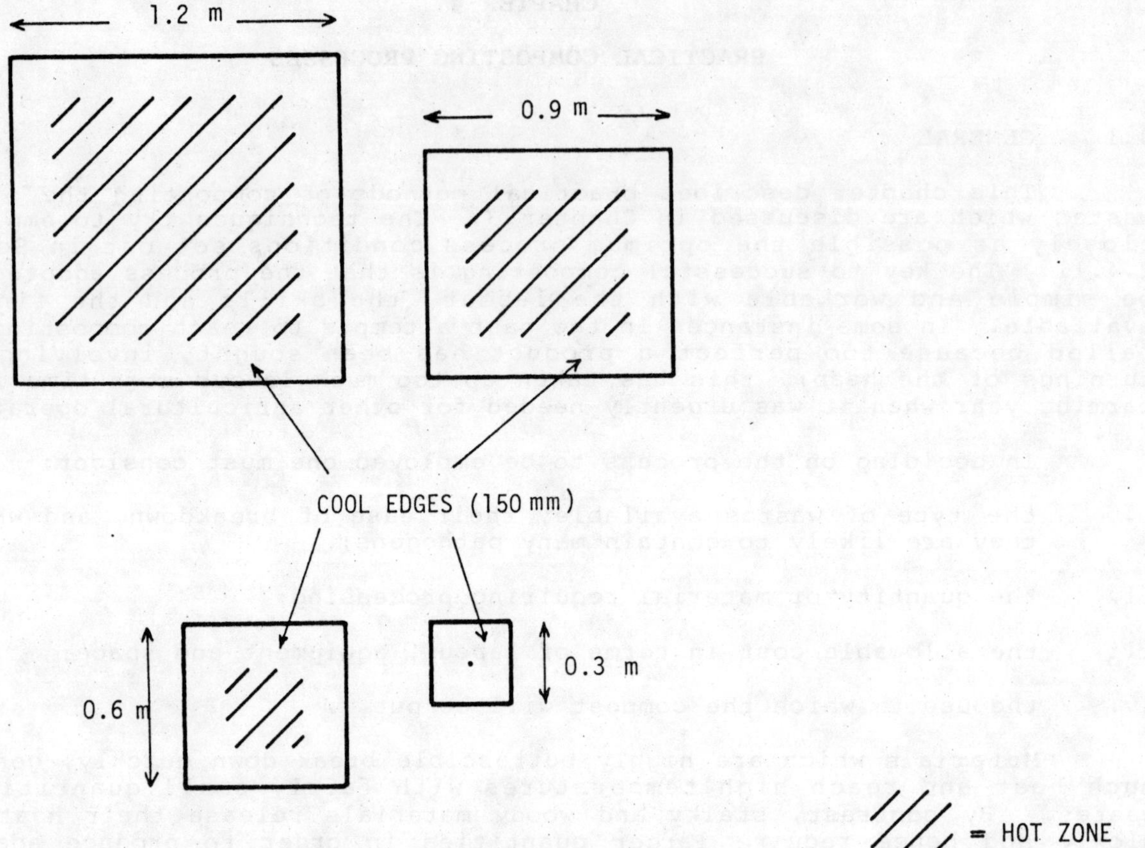

Figure 15 Effect of heap size on hot zone

Table 12 EFFECT OF COOL EDGE 150 mm THICK ON HOT ZONE OF A COMPOST HEAP

Cube edge m	Cube volume m^3	Surface area m^2	Area per unit volume m^2/m^3	Core edge m	Core volume m^3	Core volume as % of cube volume
3.0	27.000	54.00	2.0	2.7	19.683	72.9
1.5	3.375	13.50	4.0	1.2	1.728	51.2
1.2	1.728	8.64	5.0	0.9	0.729	42.2
0.9	0.729	4.86	6.7	0.6	0.216	29.6
0.6	0.216	2.16	10.0	0.3	0.027	12.5
0.3	0.027	0.54	20.0	0.0	0.000	0.0

At certain times of the agricultural year labour is fully employed on tillage, sowing or harvesting operations. At other periods there may be no work. Composting operations should be timed to fit in with slack periods. Householders and small farmers will not usually be able to afford special equipment for composting, other than their normal agricultural tools; larger farmers, towns and cities should be able to purchase increasingly mechanized equipment to help with processing.

The time taken to produce compost will depend to some extent on the types of wastes being handled. Fresh green material will break down quickly but will have little residual value in the soil as it will be oxidized quickly. Tough woody fragments will decompose slowly in the heap and in the soil afterwards so that they will exert a beneficial effect on soil properties, especially water retention, for several years. For best effect in the tropics and subtropics the compost should contain a high proportion of these small woody fragments. With the ambient air temperatures of these zones, and using the techniques described in this chapter, immature compost should be ready in 8 weeks and mature compost in about 16 weeks.

The use to which the compost is to be put will govern the degree of maturity required and hence the length of time of the composting process. When used with fine seeds the compost needs to be highly mature with no acids or ammonia being generated. For mulching round established trees and shrubs and between growing crops, relatively immature compost is suitable; this will continue to break down on the surface and be drawn into the earth by the small soil animals such as ants, termites and worms.

4.2 SMALL HEAPS

A small compost heap either in a stack above ground or in a pit may be defined as one containing approximately one cubic metre of material weighing approximately half a tonne. As shown in Table 12, in an unprotected heap of this size possibly only a third of the material in the central core will reach a reasonably high temperature in temperate climates. Under tropical conditions the hot zone will naturally be greater.

4.2.1 Structures

For successful composting on this scale the main requirements are outer wall protection, aeration from underneath, a heat insulating blanket on top of the wastes and overhead protection against rain. These are illustrated in Figure 16.

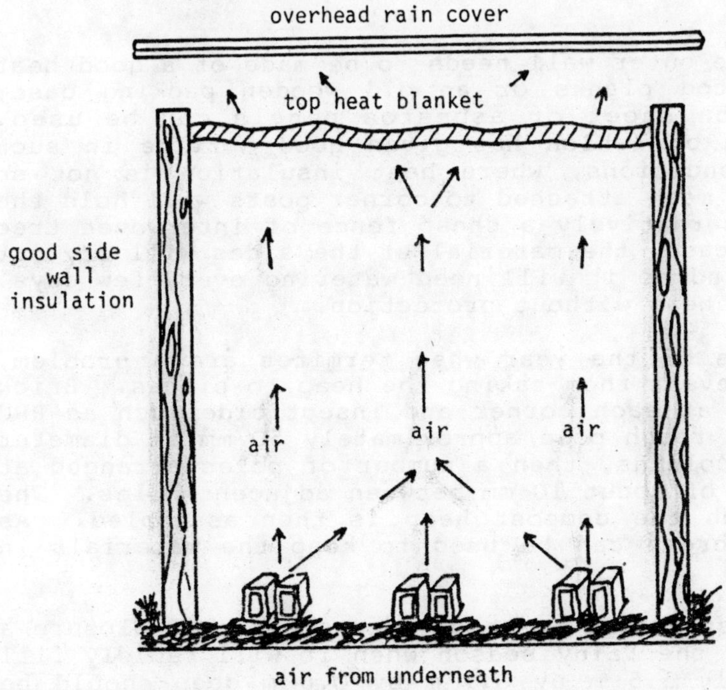

Figure 16
Essentials of a small compost bin

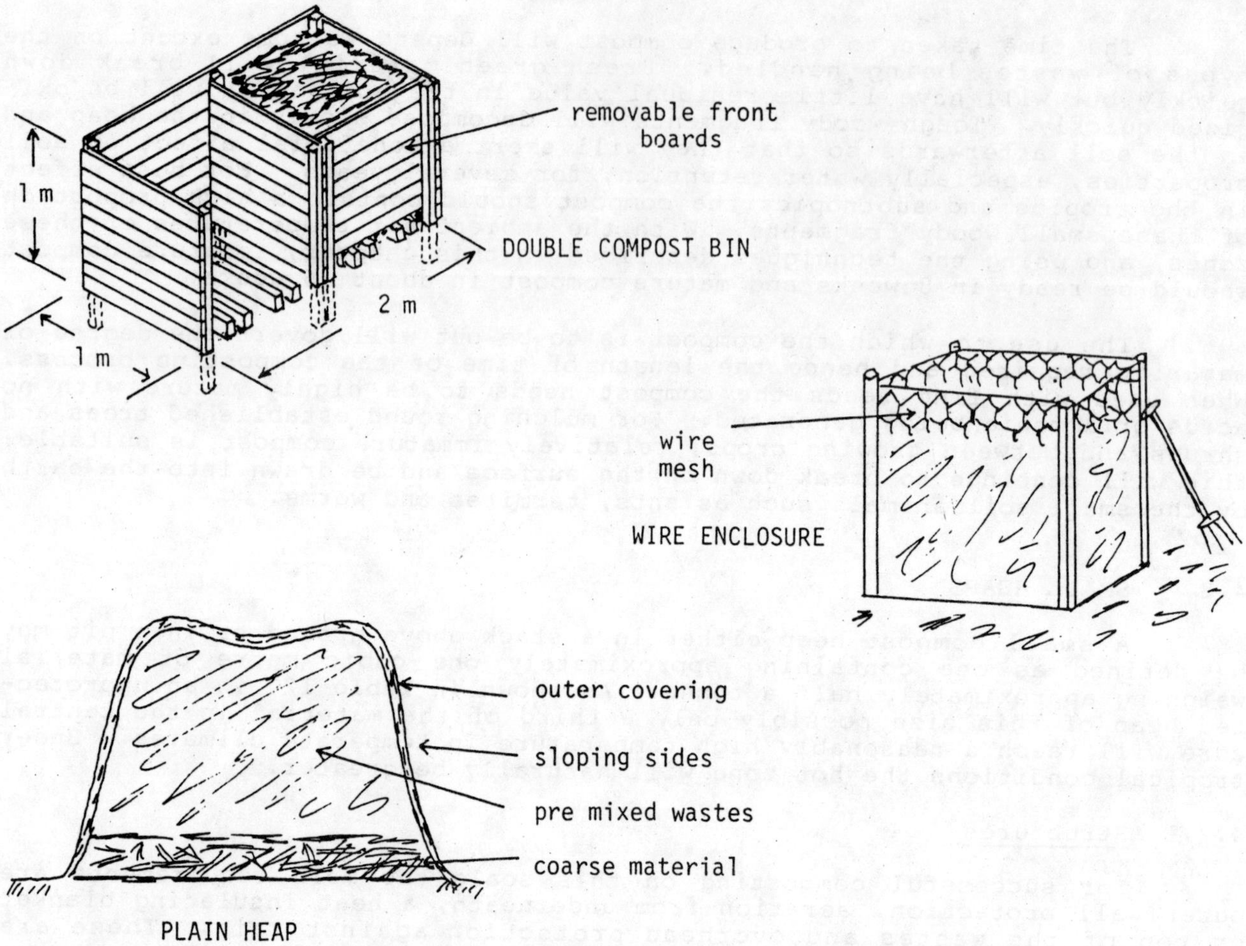

Figure 17 Double compost bin, mesh enclosure and plain heap

In temperate climates the outer wall needs to be made of a good heat insulating material such as wood planks or an old wooden packing case; failing these, corrugated iron sheet or asbestos panels can be used. Figure 17 illustrates a wooden box which has given good service in such conditions. Under tropical conditions, where heat insulation is not so important, wire mesh or plastic mesh attached to corner posts will hold the wastes together in a heap; alternatively a cheap fence of interwoven tree branches can be made. In this case, the material at the sides will dry out with exposure to the sun and wind so it will need watering every few days. Figure 17 shows a free standing heap without protection.

In hot climates at times of the year when termites are a problem, precautions must be taken to prevent them taking the heap to pieces. Brick columns 150 mm high are set up at each corner and insecticide such as BHC powder sprinkled round them; a rough pole approximately 50 mm in diameter is placed across each pair of columns, then a number of poles arranged at right angles, with a narrow gap of about 10 mm between adjacent poles. The whole forms a platform on which the compost heap is then assembled. As before, side walling or mesh screen can be used to keep the materials in place.

As an alternative to constructing an above-ground box or enclosure a pit may be used, but not during the rainy season when it will rapidly fill with water. A pit approximately 1.5 m by 0.9 m by 0.6 m deep should be ample to hold one cubic metre of composting wastes.

Compost heaps and pits should be sited away from, and down wind of, the house and preferably in a shady sheltered place to give protection from sun and wind. Ideally they should be situated where they can be inspected daily to ensure that animals, vermin and flies are not causing problems.

4.2.2 Aeration

Air should be able to circulate freely underneath the composting mass. The air filters upwards into the heap and is then warmed in the heap, becoming less dense and more buoyant, and rising up through the mass, pulling in more fresh air from underneath. This is the 'chimney effect' similar to that obtained with coal or wood fires and chimneys. It ensures that as long as the compost heap is at a higher temperature than the surrounding air, there will always be a flow of air upwards through the heap so that it stays aerated and aerobic.

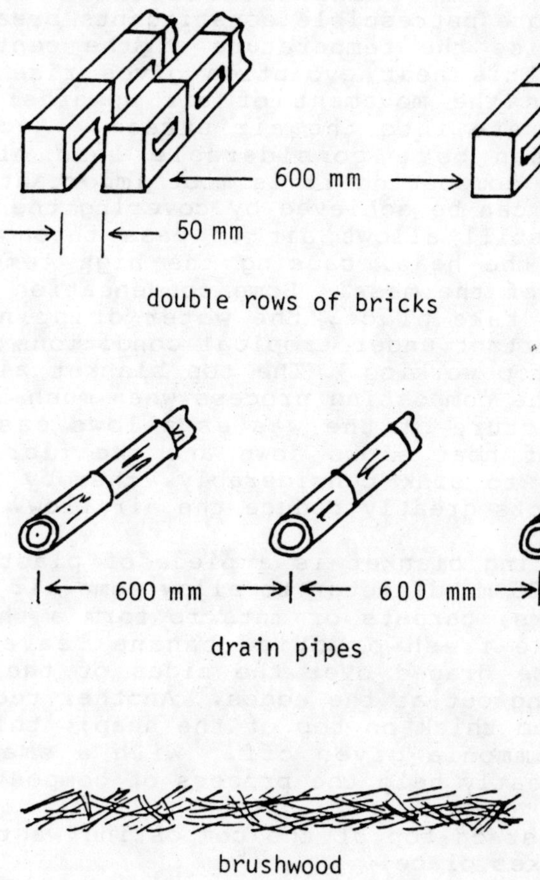

Figure 18 Methods of aeration

With the compost mass built on the anti-termite platform described above, the air flow is ensured by arranging narrow gaps between the wood poles underneath. When the heap is built directly on the ground, aeration is provided by one or two rows of double bricks with 50 mm between bricks and 600 mm between rows, as shown in Figure 18. The bricks stick out slightly from the sides of the heap and are covered with twigs or stalky material to prevent fine particles falling into the air duct and blocking it. An alternative to bricks is the use of drain pipes about 50 mm in diameter with 600 mm between rows as shown in Figure 18. Another method is to use a layer 75 mm thick of brushwood, thick hedge clippings or tree prunings under the heap; these will not break down very much and most can be re-used in subsequent compost heaps. A further technique is to dig a pair of shallow trenches about 600 mm apart and cover them with wire mesh. The trenches should be about 150 mm wide by 100 mm deep. The compost heap is then built over these aeration trenches.

Where the wastes are confined in a pit, the layer of brushwood is the best technique; however, air must be given access to this base layer. This is done by using a vertical stake, about 75 mm in diameter driven into the centre of the pit bottom, around which the heap is assembled; on completion of the heap the stake is withdrawn, leaving a vertical air passage.

4.2.3 Insulating Blanket

At the start of the composting process there is a considerable evolution of heat as the more putrescible constituents break down; this is normally sufficient to raise the temperature of the central core of the heap to at least 60°C. This heat evolution gives rise to the 'chimney effect', setting in motion the movement of air upwards; the heat also causes moisture to evaporate into the air stream. Accordingly, in an unprotected heap there can be a considerable loss of both heat and moisture. To achieve good composting it is most important to reduce these losses to a minimum. This can be achieved by covering the heap with a heat insulating blanket which still allows air to pass through. As a result more heat is retained in the heap, causing the high temperature zone to move out towards the edge of the mass. Some condensation of moisture from the warmed air will also take place, the water dripping back into the wastes. This is very important under tropical conditions as compost heaps can quickly dry out and stop working. The top blanket also regulates the air flow at the start of the composting process when much heat is liberated and the fairly open structure of the wastes allows easy air movement. Later on, the evolution of heat slows down and the fibres in the wastes weaken and allow the mass to sink considerably, thereby closing down the air pores; these two effects greatly reduce the air flow.

One form of insulating blanket is a piece of plastic sheet pierced with a few holes of about 25 mm diameter to allow some air to pass through; on top are placed old sacks, carpets or mats to form a thickness of up to 25 mm. Other methods use fresh or dried banana leaves or woven palm leaves; these can also be draped over the sides of the compost heap to prevent heat loss and drying out at the edges. Another technique is to put a layer of soil 25 to 50 mm thick on top of the heap; this will also help to trap any smells and ammonia given off. With a small heap such an insulating blanket will greatly help the process of composting.

The blanket must rest on top of the composting wastes and sink down with them as settlement takes place.

4.2.4 Overhead Protection

Compost heaps work much better if overhead protection against the weather is provided. Heavy rain falling onto an unprotected heap can make it too wet or even waterlogged, fill up the air spaces, cause loss of heat and leach out valuable nutrients into the soil underneath. Very hot sun and strong wind will cause a heap to dry out quickly; this is one reason for using compost pits wherever possible in the tropics.

The overhead cover can be made from a piece of corrugated iron sheet or coconut palm thatching. It should be mounted preferably at an angle, at least 150 mm above the heap and be removable or arranged so that work can be done on the heap.

4.2.5 Assembling the Heap - Ample Wastes Available

Small compost heaps are usually associated with a household and garden. There is normally a fairly even flow of organic material coming from the house, such as kitchen wastes and night soil. Debris from the garden varies in quantity through the year, reaching a peak just after the rainy season in the tropics and falling to a smaller quantity in the dry season. In temperate climates there are very few compostable garden wastes during the winter months.

With small heaps the ideal situation is to have ample wastes available so that the compost box, heap or pit can be filled on the same day. A rapid evolution of heat then takes place so that the high temperature spreads out close to the edges of the heap; after about a week the heap will have sunk to about half its original height but will still be hot. Further wastes can then be added to top it up, thereby using the space most economically.

To construct a small compost heap the underneath air supply is first assembled as discussed in Section 4.2.2. Then the compost box or container is erected. Next the kitchen and garden wastes are brought together to start building the heap. Approximately two small barrow loads of material, totalling some 75 kg, are assembled at a time on the ground beside the container or pit. If the wastes contain a lot of straws and stalks, their high C/N ratio should be balanced if possible by adding an activator with a low C/N value, such as manure or fresh green materials, as shown in Table 5. Any tough stalks and stems and other large pieces of material should be broken up. To the mass is then added a spadeful of soil, about 1 kg, some wood ash if available, and ideally about 1 kg of product compost from an earlier heap. The material is now mixed together, until the composition is reasonably uniform. If the mixture appears to be too dry then it is sprinkled with water until it is moist but not sodden. A good guide to follow is that the material is sufficiently moist if the surface of the particles glisten. This pre-mixing is most important to the proper working of the compost heap. The batch of wastes is then lifted into the compost box and levelled off roughly; it should form a layer approximately 200 mm deep.

The procedure is repeated with another two barrow loads of wastes, till the 200 mm thick layers in the box or pit reach a height of about 1 m. It is then levelled off and the central aeration stake removed from the pit. Finally the top insulating blanket and the outer cover for rain protection are put in place.

Where night soil is used, some is spread on each layer of wastes, but it is kept away from the edges of the heap so that flies cannot have easy access.

If the heap has been properly made it should warm up quickly and reach a peak temperature at the centre of at least 60°C within 7 days. Steam should be apparent if the top insulating blanket is peeled back. If the compost has been made in a wooden box with good side wall insulation, the heat should extend out from the centre to give a temperature of approximately 50°C about 75 mm in from the walls. By this time the whole mass should have settled down to about two-thirds of its initial height. At this point the heap can be topped up by adding several more layers of the pre-mixed wastes.

4.2.6 Assembling the Heap - Insufficient Wastes Available

With small households and gardens insufficient wastes may be generated to fill a box or pit of 1 m^{3} size in one operation. The consequent practice of adding small daily or weekly quantities to a cold compost heap does not allow generation of sufficient heat to ensure an adequate kill of weeds, seeds or pathogens. This is a serious problem for many households, but can be solved in a number of ways:

i. the box or pit is divided with a vertical partition; one of the two compartments is built up at a time. A tall pile with small cross-sectional area is more effective at composting than a shallow heap over a large area;

ii. extra wastes can be brought from off the premises - from neighbours, markets and forests - in order to make up the required amount of material;

iii. communal heaps can be built using the wastes from several neighbours;

iv. a long-term storage method can be used. In this technique the compost box or pit is used first as a storage container. The organic waste is tipped into the box as it becomes available and is kept as cool and dry as possible to reduce decomposition. It is neither mixed nor moistened and the underneath air supply is not installed. The heat insulating blanket is not laid over the mass but the overhead rain protection must be put in place. Care must be taken with putrescible wastes to prevent them giving rise to fly breeding; the mass can be covered with a piece of plastic sheet or a layer of soil. The material will certainly go anaerobic and cause smells but the soil or sheet of plastic over the top will prevent these odours spreading too far. When the box or pit is full the wastes are then emptied out and used to build a proper compost heap. This is done in the manner described in the previous section. Because some decomposition will have taken place during storage, a high energy activator will be necessary to get the heap to warm up quickly, such as young green grass cuttings, manure slurry, biogas sludge or urine. Some of the activator is mixed in with each batch of wastes prior to loading them into the heap.

4.2.7 <u>Turning the Heap</u>

The need to turn, or not to turn, the material in a compost heap is probably the process factor which decides whether composting is an acceptable practice in a community.

As shown in Section 2.4.7 agitation or turning has a definite beneficial effect on the composting process. It helps mix the various wastes in the heap; it moves the cool and probably dry outer edges of the heap to the hot centre; it aids aeration where air has had difficulty in penetrating to the middle of the heap; it also gives an opportunity to moisten the material if this has become too dry. In a well-mechanized factory for composting urban wastes a lot of agitation is normally done; as a result the wastes can be brought through the mesophilic, thermophilic and cooling down stages in about 7-10 days; the maturing stage takes much longer and is probably not greatly affected by the amount of agitation. In a similar way, by turning it every few days, a small heap of garden and household wastes will undergo considerable breakdown in about 14 days, especially in the tropics.

However, in an un-mechanized composting system turning requires the expenditure of a significant amount of human energy; this takes time and costs money. It is normally a job for the young and physically fit who have few other tasks to do. However, in many parts of the world where the recycling of organic wastes is urgently needed, such labour conditions are frequently not available. Where the women in the household are responsible for growing the food, they have many other tasks to perform. Hence it is vital to the widespread uptake of composting that the chore of heap turning be greatly reduced or, better still, eliminated. The need to speed up the composting process from taking 2-3 months down to taking 2-3 weeks is rarely of vital importance in household and farm situations.

Some of the early enthusiastic advocates of composting recommended turning a heap or pit three times. This was later reduced as more experience was gained and the physical burden of turning was appreciated. The present authors consider that if good care is taken in assembling the heap, turning can normally be omitted. This initial care includes use of a compost box with side wall insulation and an air supply from underneath, construction of the heap from pre-mixed wastes in one operation although topping-up a few days later is allowable, and the presence of a heat insulating blanket on top of the wastes. Using this approach one of the present authors (KRG) has made annual batches of garden compost in converted wooden packing cases for some 20 years; the produce has proved satisfactory without the necessity for turning.

Where there is no side wall insulation the outside edges of the heap will be cool, probably dry, and will not have composted properly. When the heap is dismantled these outer edges can be cut off, leaving the properly composted inner core; the uncomposted edges are then used as part of the next compost heap.

Where the wastes are simply piled on the ground without provision of an air supply, or fast breakdown is required, the heap can be turned. This should be done about 10-14 days after the heap is first built, while the heap is still hot and the wastes have enough energy to build up a high temperature again. The sides and top of the original heap are turned into the middle of the new heap and covered with the remaining material.

4.2.8 Monitoring the Process

The progress of the composting reaction can be followed approximately with a few simple tests. If a heap contains quite a lot of fresh green plant material it should warm up to peak temperature in about three days; where it contains much strawy, stalky material plus manure it will take possibly seven days to reach the same temperature. This evolution of heat will cause air to start circulating through the heap and moisture to rise to the top of the wastes.

Progress should be examined 3-7 days after heap construction. The top insulating blanket of plastic sheet and sacking, or banana leaves, is peeled back until the centre of the top of the heap is exposed. There should be a definite feeling of warm rising air and the underside of the blanket and the top of the wastes should be hot. The underside of the blanket should also be wet with drops of moisture. A metal or wooden rod can be pushed 300-600 mm into the centre of the heap for 10 minutes; on withdrawal it should be definitely hot and often too hot to hold. In addition the rod should be moist, possibly with steam rising from it. If temperature and moisture are following this behaviour the composting reaction is satisfactory. The wastes should also have settled down in height quite noticeably. If the temperature is satisfactory extra layers of pre-mixed wastes can be added, building the heap back to its original height.

The heap can be re-examined about 14 days later, 17-21 days after it was first built. With a small heap the temperature peak should have been passed, although the wastes should still be warm. The underside of the top blanket should now be dry because the rising air should no longer contain excess moisture.

Four weeks after construction the heap should be cool and the small soil animals such as manure worms and mites will probably invade the mass and help the micro-organisms with the task of decomposition.

Provided that an adequate air supply has been arranged underneath the compost heap, there should be no bad odours given off once the heap has reached peak temperature. Even if anaerobic smelly wastes have been incorporated into the heap the movement of air as the mass warms up should soon stop the odour.

Composting processes occasionally go wrong. If the temperature does not rise properly in the first few days, the pile may be too dry, too wet or the C/N ratio too high. If there is too little moisture for the micro-organisms, little heat is generated and the underneath of the top blanket will be dry. A sample of the wastes should then be removed from the interior of the heap; this will probably look dry and on squeezing gently by hand will not exude drops of moisture. In this case the heap will have to be rebuilt with more water sprinkled on to the wastes. A good test for adequate moisture is that on squeezing a handful gently moisture droplets are given out.

If the wastes are too wet, the air passages will probably be filled with moisture and the mass be anaerobic with a foul odour. A sample from the interior will drip moisture with very little squeezing. The remedy is to rebuild the heap, incorporating more absorbent strawy material to soak up the excess moisture.

If the C/N ratio is too high, particularly if the heap contains an excess of fairly unreactive woody wastes, more fresh green plant material or manures should be added to increase the energy release. If such materials are not readily available, the heap will need to be made much larger to increase the quantity of wastes undergoing breakdown.

Where the mass has a lower than optimum C/N ratio no problems should arise as long as the heap is readily permeable to air. The temperature should increase easily but a slight odour of ammonia will develop; this is not serious for the process but represents a loss of the plant nutrient nitrogen.

4.2.9 Maturing the Product

About 4 to 6 weeks after its construction a compost heap will have cooled back to the temperature of the ambient air. Most of the breakdown process will have taken place and the majority of the air requirement supplied. The weight of the wastes will have fallen to about half of the original amount due to oxidation of much carbon to carbon dioxide and loss of moisture. The volume will have fallen to nearly one third with the loss of weight plus the breakdown of larger particles to smaller ones which pack together more closely. The colour will have changed to a dark brown-black. Maturing now takes place in which the broken-down fragments of the wastes are slowly converted by polymerization into the very complex and stable humus product called compost. This appears to be mainly ligno-protein, formed by combination of lignin residues with microbial protein.

The amount of maturity required in a compost depends on the use to which the product will be put. For mulching on the surface of the ground between rows of crops or around established trees and shrubs, immature compost is perfectly acceptable. Its further maturing then takes place on the surface and the fragments of organic matter are eventually drawn down into the earth by the small soil animals such as worms, or else incorporated by subsequent tillage operations.

Where the compost is required for direct incorporation into the soil, especially prior to sowing small seeds, a high degree of maturity is required. It is essential for good seed germination and seedling growth that the compost no longer release organic acids, or ammonia, or rob nitrogen from the soil water in competition with the crop. Such robbing can occur when wastes of high C/N ratio are composted and the product has not yet reached a C/N ratio below 15/1 before use.

Since the maturing stage makes little demand on air supply and none on heat conservation, it does not make economic sense to retain the compost in a box or pit, especially if further raw wastes are waiting to be processed. The immature compost is dug out from the box, heap or pit and stacked on the ground; great care should be taken to cover it up with a plastic sheet or banana leaves to prevent rain leaching out the plant nutrients, and also to safeguard it from termite attack where this is a problem.

4.3 LARGE HEAPS

Large compost heaps contain at least 0.5 tonne of material and possibly up to 100 tonnes when in the form of an extended pile (windrow). They may be sunk in the ground in pits or be above ground; this often depends on the height of the water table in the ground and whether it is the rainy season. The use of pits in the dry season provides protection against heat loss and drying out of the material. Although pits have to be dug out initially, it is easier to form heaps in them than in stacks because there is less lifting up of material.

The quantities of waste material needed for a large heap come from farms, plantations, communal agricultural enterprises or villages and towns treating their daily production of refuse and night soil/sewage sludge.

There are several well-tested forms of large heap system which can be operated by manual labour. These are the Indore system for pits and stacks originated by Sir Albert Howard in India in the ninteen-twenties, the high temperature stack used by the Chinese since about 1960, the Bangalore aerobic/anaerobic pit method worked out in India in the ninteen-thirties, and the Mazibuko or 'fertility trench' system developed in Africa about 1956 where composting is combined with hillside erosion control.

Due to the greater volume of material in a large heap, more of the wastes will be exposed to the central core temperature than is shown in Table 12 and Figure 15 for a small heap. In an above-ground compost stack having the dimensions given in Figure 19, assuming a cool edge 150 mm thick, the hot core volume will be 68 percent of the total volume. This value is increased by using a pit, or by covering the stack in an insulating layer, Nevertheless, the aeration and temperature throughout an unagitated compost heap can never be completely uniform. As shown in Figure 20, for an aboveground stack there is a cool outer skin and very cool lower edges; the hottest part is just above the mid-point while there is often an anaerobic zone just above the base if aeration depends on diffusion from outside the heap.

In any composting system employing natural aeration there are nearly always periods when anaerobic conditions are present. This is particularly true during the first few days after a heap is built, when the more putrescible components are breaking down and the need for oxygen is greater than the ventilation can supply air. This does not appear to cause problems as long as some air is moving up through the heap. The condition gradually improves once the peak temperature has been reached and the more resistant components come under attack; oxygen demand and supply are then

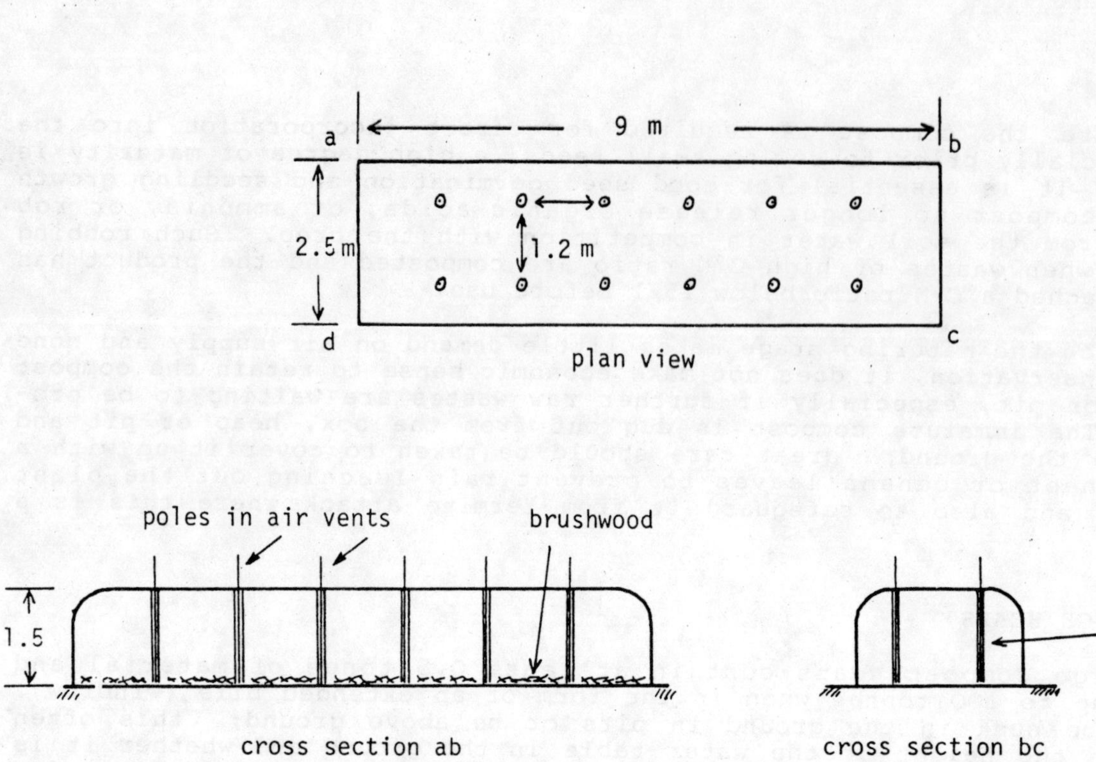

Figure 19 Typical compost stack above ground

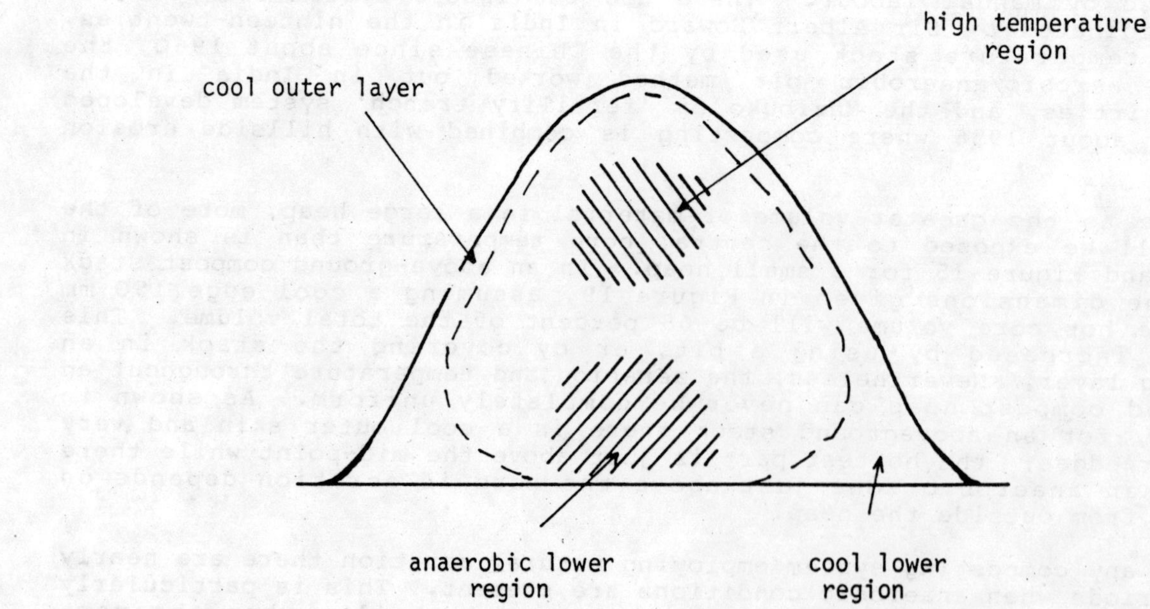

Figure 20 Variation of conditions in a compost heap

more evenly matched. Compost is very effective at absorbing odours provided that these are not excessive. Odour problems only really arise from anaerobic conditions when the air supply is reduced to virtually nil, either through completely inadequate ventilation arrangements or because the moisture content is so high that the heap is waterlogged.

4.3.1 The Indore Heap

In this method compost is produced in pits during the dry part of the year and in above-ground stacks, between the pits, in the rainy season.

i. **Dimensions and siting of the heap**. A heap (pit or stack) built to a height of 1.5 m and having a base area of about 40 m^2 will require nearly 20 tonnes of organic wastes, about 80 m^3, and should produce 6-8 tonnes of compost per batch. It should be possible to produce 2-3 batches per year from a pit and normally one batch from a stack. One or more pits may be necessary to handle the volume of the wastes requiring treatment.

Typical dimensions of a pit are given in Figure 21 while Plate 9 shows a pit under construction. The pit is excavated to a depth of 600 mm and has sides sloping at 45° as shown. The excavated soil is used to build an earth bund around the pit to a height of 300 mm, giving a total depth of just under one metre. Important points to note are:

a. the pit should be sited well away from any source of drinking water, a distance of at least 25 m being desirable;

b. it should be constructed with its long side at right angles to the prevailing wind direction;

c. use should be made of existing wind breaks, such as buildings or trees, to provide shelter from wind, sun and rain;

Plate 9 Digging a compost pit. The excavated soil is used to raise the height of the sloping edge
Source: FAO filmstrip Compost - Thailand

ii.

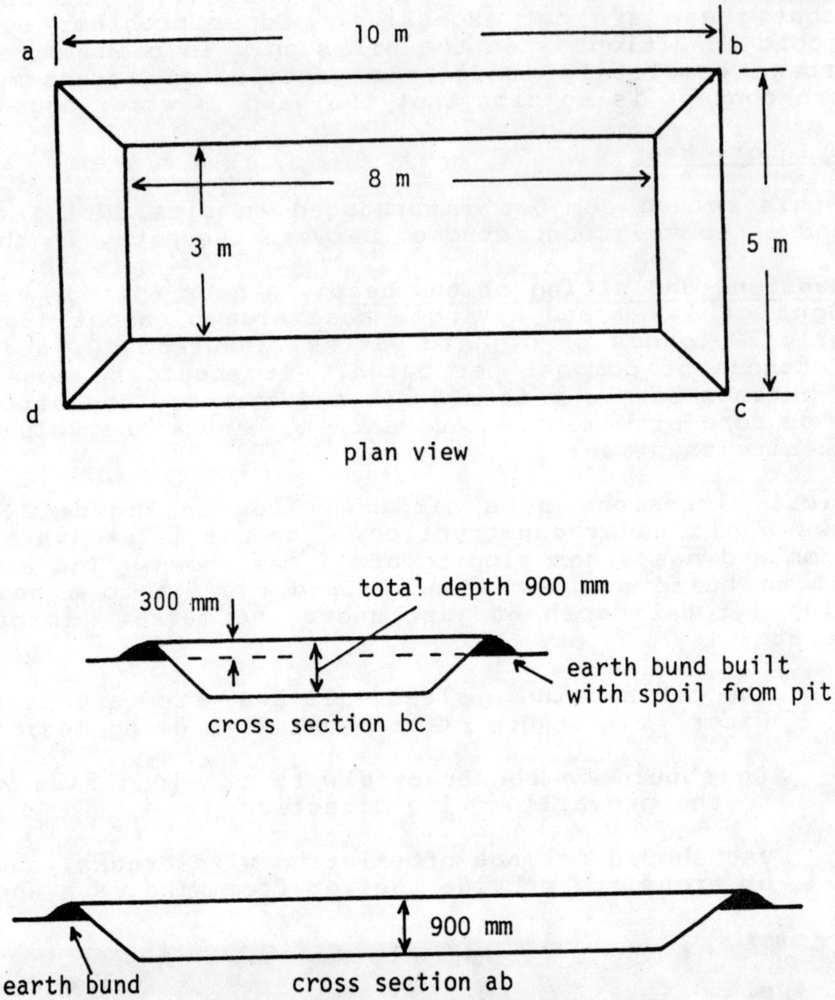

Dimensions:	length	breadth
bottom of pit	8 m	3 m
ground level	9.3 m	4.3 m
top of bund	10 m	5 m

Figure 21 Details of a compost pit

d. if natural shelter is not available then a light and simple roof should be constructed over the heap; this will improve the process by reducing moisture loss through sun and wind in the hot season and restrict loss through leaching by rain in the rainy season. A frame of rough poles strong enough to withstand heavy rain should provide sufficient support. The roof should be sloped and have a minimum height above the heap of 50 mm to allow aeration. The covering can be made from local material such as coconut matting, grass matting, old fertilizer bags or plastic sheeting. If a non-porous cover is laid on top of the heap then it should be perforated with 10 mm diameter holes every 150 mm to aid aeration;

e. sufficient space should be left between pits, at least 4 m, to enable compost stacks to be built during the rainy season.

Forming the heap. The heap will be assembled from the types of organic wastes discussed in Chapter 3 which are available in the locality at the time of year. Ideally they should include some fresh green material, some straws and stalks, woody materials and manures or night soil. Materials should be stored until sufficient are available to make one or more heaps.

When building a heap, whether in a pit or a stack, it is a good idea to start by laying old branches or brushwood at the bottom to a depth of about 150 mm. This will help with aeration and prevent waterlogging of the lower layers. The base area of a pit should be divided into six roughly equal sections, five of which are filled and the sixth left vacant, as shown in Figure 22. Building is carried out one section at a time, beginning in the first section. Each section is made up of 7 to 10 layers, approximately 230 mm deep, giving a total constructed height of 1.5 m after sinkage has occurred. Within each layer quantities of the various compostable raw materials are incorporated in proportion to the total amounts available for the complete operation. For instance, in a 230 mm layer one may have 100 mm of coarse dry wastes, 75 mm of green weeds and leaves, 50 mm of manure and a sprinkling of urine-earth/wood ash. In this way no ingredient will be exhausted before the heap is complete. Materials can be carried to the heap in baskets or stretchers, emptied on and raked level (Plates 10 and 11). Each layer is moistened with water until it is damp but not sodden (Plate 12). In the winter and early spring when ambient temperatures are lower and moisture evaporates from the heap more slowly, rather less water need be used; in the summer and autumn with high ambient temperatures and evaporation rates, more water can be added. The layering process is shown in Figure 23 which depicts a partly-built heap.

Restricting the maximum depth of any plant ingredient to 100 mm and any manure to 50 mm will assist in aeration. Materials with very poor aeration qualities, such as grass clippings and sawdust, should not be placed in layers but should be premixed outside the heap with a coarse ingredient.

When two layers of a section have been completed, ventilation holes should be started. This is done by probing vertically through the wastes with a pole and gradually increasing the size of the hole to approximately 100 mm diameter by waggling the pole. The pole is left in position to ensure that the air vent continues to the top of the completed section. The first hole should be at a distance of not more than 600 mm from the edge of the heap and the maximum distance between the holes should be 1.2 m. For a wide pit three ventilation holes per section will be necessary.

Once the air vents have been established, the building of the section can continue as before until a height of 1.5 m has been reached. Building of the second section can then begin and so on until five sections have been completed. The sixth section is left empty, being used later when turning the heap (Figure 22). It is very important not to walk on the heap at any stage as this will compress it and decrease aeration.

The heap should be lightly watered on completion and again the next morning. A diagram of a compost stack above ground is given in Figure 19. This shows the ventilation poles, which are left in place to prevent the air vents collapsing, and the height of the heap above ground level.

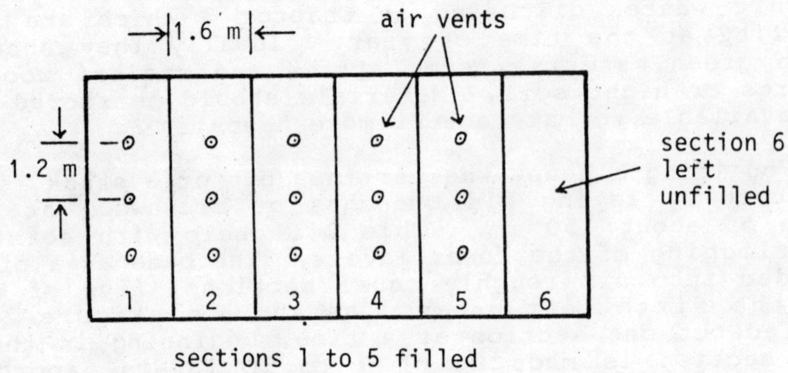

Figure 22 Plan view showing completed compost heap

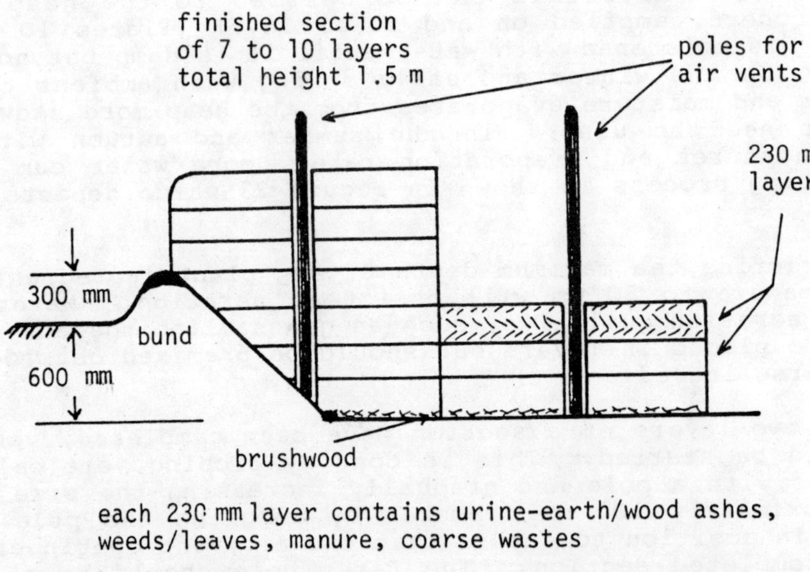

Figure 23 Partly built pit compost heap showing layers of ingredients

iii. **Turning the heap**. Within a week of completing the compost heap the centre should have warmed up to its maximum temperature; a little steam should be rising out of the vertical air vents. The combined effect of heat, moisture and attack by the micro-organisms weakens the fibrous strength of the organic waste and the mass sinks to the level of the bund wall around the heap.
The heat extends through the heap to about 150 mm from the sides. The temperature falls rapidly in the outer zone, reaching ambient level at the earth walls of the pit or the exposed edge of an unprotected heap. The material in this cooler region composts slowly and the weeds, seeds, and diseased material may not be completely destroyed. Fly larvae and pupae can develop in these cooler areas but will be destroyed by the higher temperatures in the centre of the heap. Hence it is most important to turn the heap so that the

Plate 10 Stretcher for carrying wastes to the compost pit
Source: FAO filmstrip Compost - Thailand

Plate 11 Tipping wastes into a pit; they are spread out into an even layer
Source: FAO filmstrip Compost - Thailand

Plate 12 Watering wastes in a pit; they are watered evenly as each layer is completed
Source: FAO filmstrip Compost - Thailand

Plate 13 Completed heap ready for turning about 10 to 14 days after building. This speeds up the process by extra aeration and mixing. A second and even a third turn may be desirable in special circumstances
Source: FAO Filmstrip Compost - Africa

cooler top, bottom and sides are mixed back into the middle of the new heap once the central temperature has started to drop. Turning also helps the thorough mixing of the ingredients in the heap and enables the air and water situation to be checked. Plate 13 shows a heap ready for turning.

Normally the first turn is carried out ten to fourteen days after building the heap. Up to this time the only attention necessary is the occasional shaking of the poles in the air vents. Turning is done by putting the top 150 mm from section five on to the floor of section six, mixing it in the process. Water is sprayed on as necessary. The next 150 mm is then put on top of this and watered. The air vents must be re-established. When section five has been completely turned into section six, section four can be turned into section five and so on until each section has been turned, leaving section one vacant. The first turn is now complete. The mass should warm up again and some steam be evolved. No further attention is required except for keeping the air vent holes open.

The second turn is carried out after a further three weeks, during week five after heap construction. If the heap has dried out, or the original materials are still recognizable after nine weeks, a third turn is recommended.

The final compost product should be mature about 12-16 weeks after starting, although it may take longer in the rainy season. The mass will have cooled down and shrunk to less than half its original volume. Apart from some small portions of twigs, none of the original components should be recognizable. Mature compost should have a crumbly texture, an earthy smell and be dark brown or black in colour. If it is required to improve the appearance of the compost the material can be thrown against a piece of wire mesh screen sloping back a little from the vertical. Fine material passes through the screen while oversize pieces tumble down the front and can be recycled through another compost heap. For success in screening the material needs to be fairly dry, below 40 percent moisture content on a fresh weight basis.

There is controversy about the necessity for turning the compost heap as it is a labour consuming operation. Insistence upon frequent turning has probably prevented acceptance of the composting process for agricultural wastes, hence considerable common sense and judgement is needed. Turning undoubtedly has a beneficial effect on composting in improved mixing of ingredients, better aeration, speeding up the process and giving an opportunity for moisture adjustment. When the time is short between harvesting one crop and preparing the ground for its successor, turning the heap 2-3 times could make all the difference between having fairly mature compost ready for spreading and having only partly decayed material still in the heap. However, when there is no such urgency and particularly where farm labour is being fully occupied in more important tasks, then turning can be reduced. At the worst, with no turning, the less broken down cool zone on the top and sides of a heap can be removed and used for a subsequent heap; the inner core which should have passed through a satisfactory temperature-time cycle is ready to be used. The authors advocate the use of at least one turn if at all possible and especially when pathogen-bearing materials such as night soil have been included.

With any composting system it is important to keep the working site tidy. Plate 14 shows a number of above ground stacks laid out on a tea estate while Plate 15 depicts a neatly constructed long windrow in Africa.

Plate 14 Compost stacks on a tea estate laid out in a very well-organized manner to allow movement of carts between the heaps
Source: Gotaas (1956)

Plate 15 Windrow heap which has been neatly constructed and covered with a layer of soil to help retain heat and prevent fly breeding
Source: Gotaas (1956)

iv. **Monitoring the process**. With a large compost heap involving many tonnes of material some measure of control over the process is important.

For the mixture of organic wastes an initial carbon/nitrogen (C/N) ratio of 25-35/1 was recommended in Chapter 2. In practice a farmer may not have complete control over this. The composting process is fairly adaptable and can handle a wide range of organic wastes and eventually turn them into humus as explained in Chapter 2. The layering technique of the Indore method will ensure a good degree of distribution of constituents through the heap. On a farm composting scale there are no simple methods of measuring the C/N ratio of the materials or even of getting a truly representative sample to analyse.

Maintaining a reasonable moisture content in the heap is important, particularly in the dry season in the tropics when drying-out can occur. The tests mentioned in Section 4.2.8 for monitoring the progress of a small heap are mainly suitable for large heaps; the major exception is checking the presence or absence of moisture on the underside of the top insulating blanket. With large heaps such a removable blanket is rarely used. One test for adequate moisture in the wastes is by squeezing a handful gently, when moisture droplets should just be given out. Another technique is to insert into the large heap a small bundle of cereal straws, such as millet stalks, and leave them for five minutes; if the stalks are still dry on removal there is insufficient moisture; if water droplets are on the stalks the moisture level is too high, while if they are just damp then the moisture content is at approximately the correct level of 50-60 percent.

The temperature level is readily tested by using a metal or wooden rod as described in Section 4.2.8.

4.3.2 The Chinese High Temperature Stack

China is a big country with a wide range of geographical conditions and a long tradition of composting. Many techniques have been employed there using aerobic and a anaerobic conditions in handling most of the wastes described in Chapter 3. A considerable quantity of night soil is returned to agriculture, probably a much higher proportion than in any other country. In an endeavour to improve the hygienic disposal of night soil via composting, a high temperature system has been developed; this technique involves forming air channels by using bamboo poles and covering the outside of the heap with a thin layer of soil.

Figure 24 gives an indication of the arrangement of such a heap which is usually 2-3 m wide, 6-7 m long and 1-1.5 m high. A layer of crop wastes 150-300 mm thick is laid on the ground. Across this are laid bamboo poles 80-100 mm in diameter, spaced up to 2 m apart. Further bamboo poles are then arranged vertically to act as chimneys. The organic wastes - green material, straws and stalks, animal manures, night soil and often refuse - are then added in layers, as in the Indore process, until a height of 1.5 m has been reached. The heap is then completed with a covering 30-50 mm thick of mud mixed with a binder such as powdered horse manure, cinders, wheat bran or chopped rice straw. The sides of the heap need to be sloped, as shown in Figure 24, so that the mud coating stays in place.

After 24 hours the heap starts to warm up and settles slightly. The bamboo poles are then carefully withdrawn leaving the formed aeration passages and chimneys through the stack. Once the internal temperature has reached 60-70°C (after 4-5 days) the ventilation holes are often blocked. In winter months this is to reduce the effect of low air temperatures at

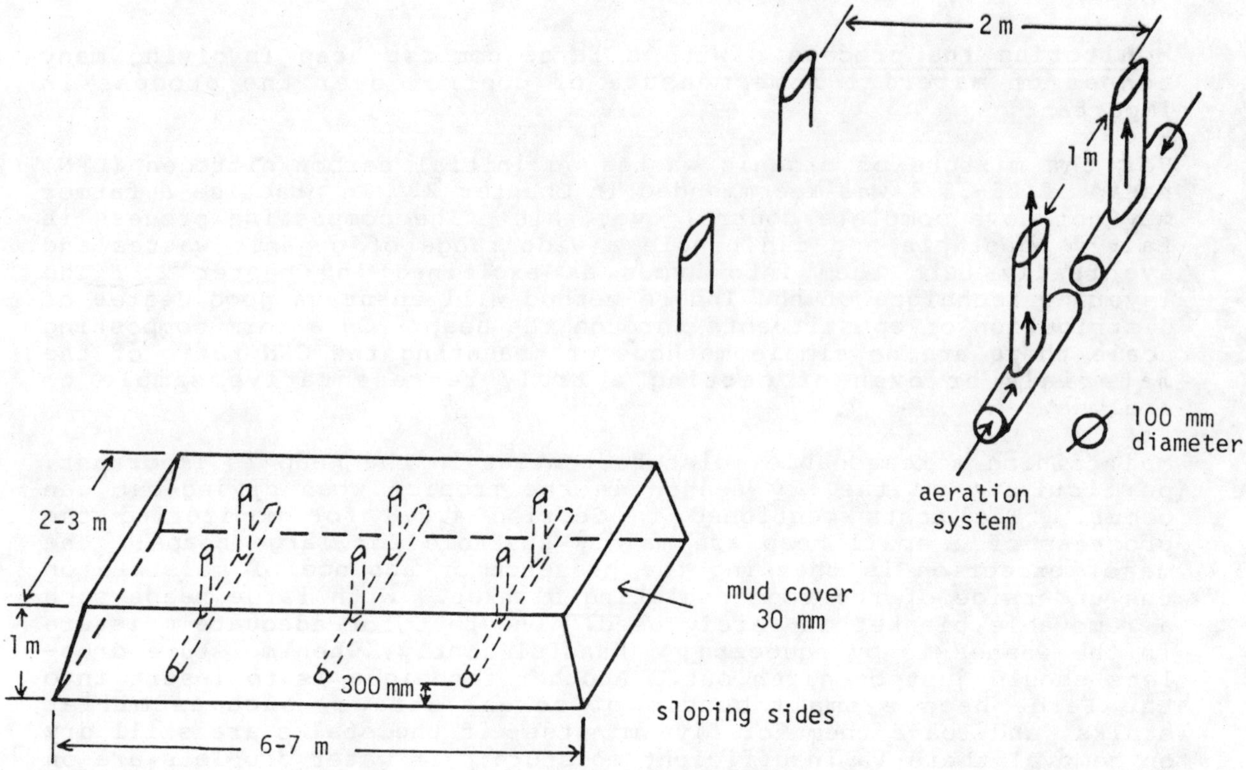

Figure 24 Chinese high temperature heap

Plate 16 Chinese high temperature heap, construction details showing bamboo poles in place to form aeration passages Source: FAO (1978c)

Plate 17 Chinese high temperature heap, mud covering in place on sloping sides; this holds in the heat and odours and prevents fly breeding Source: FAO (1978c)

night from chilling the stack; in summer it is to reduce evaporation of moisture. Plate 16 shows one of these heaps being assembled, with the bamboo poles protruding out of the sides and top of the stack; Plate 17 shows a heap being neatly plastered with its mud coating.

The heap is usually turned after 14 days to ensure good mixing of the constituents and an even breakdown. During turning, the material is restacked and water added to compensate for any major loss; the heap is then replastered with another mud coat. The compost is normally ready for use in 8 weeks; it is considered to be of high quality and free of pathogenic micro-organisms and the ova (eggs) of intestinal worms.

The mud cover of the stack helps the inner core heat to spread out to the edges, prevents flies breeding in the wastes, greatly reduces any nitrogen loss as ammonia and absorbs any bad odours. In the rainy season, the top of the stack and its mud coating need to be adequately sloped so that rain water runs off easily.

In some areas, bundles of 6-8 maize stalks are used instead of bamboo poles to form the vertical air vents for these compost stacks. An alternative method of providing an insulating cover is to use paddy straw or product compost instead of mud.

4.3.3 The Mazibuko Trench

The Mazibuko or 'fertility' trench system was developed by Robert Mazibuko in 1956 to restore soil fertility on arid, hilly areas which had suffered so badly from soil erosion due to over-grazing that little food could be grown.

The technique is to dig trenches 1.2 m deep and 2 m wide at regular intervals down the hillside while keeping to the contour as far as possible. Each trench is then filled with alternate layers of soil and organic wastes, each about 300 mm deep. In the organic layer is placed virtually all the compostable waste from the village; which includes cut grass, stems of millet and especially maize, and provides a large volume of material, animal manures and domestic wastes. The top organic layer is piled high enough to allow for sinkage of the mass over the years and then capped with soil (Figure 25).

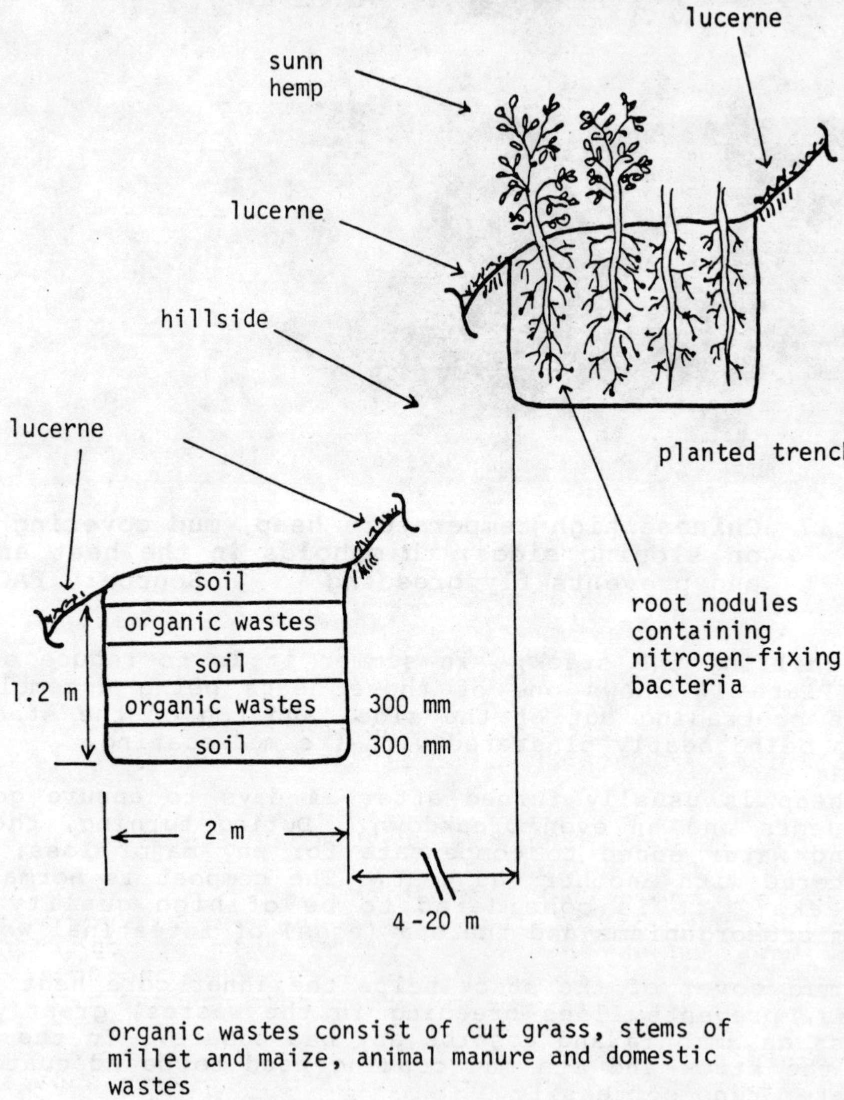

Figure 25 Mazibuko trench system

In the filled trench is first sown sunn hemp (Crotalaria juncea), a legume with deep roots which supply nitrogen to the underground compost heap, thereby helping the stalks and stems to break down. When the rains fall the water soaks into the trench which acts like a sponge. Soil erosion on the hillside is thereby stopped. The second crop is madumbe, a variety of cocoyam or taro (Colocasia esculenta) with edible leaves. This is followed by maize, millet and several types of vegetables such as carrots, pumpkins and native vegetables which can grow through the dry season, drawing on the moisture held in the 'sponge'.

The stretches of hillside between trenches are sown with lucerne which can stabilize the soil provided that the angle of the slope is not excessive.

The Mazibuko trench system combines several functions: composting the organic wastes of the village, preventing soil loss from exposed hillsides, providing a route for heavy rainfall to be channelled into the subsoil, and establishing high fertility areas with deep soils for crop growing. With further research to ascertain the best legumes to provide the nitrogen for the first crop and satisfactory breakdown of the organic wastes in the trench, the system should be widely applicable.

One particular advantage of the Mazibuko technique is that it is less affected by the activities of termites. Where they are prevalent these white ants are capable of devastating above-ground stacks of organic wastes at certain times of the year.

4.3.4 Other Composting Techniques

A simple technique employed in the Far East is to dig a trench around a tree, towards the edge of the root zone. The trench is then filled with cut grass, fallen leaves and other organic wastes and topped with soil. As with the Mazibuko system, the trench absorbs water during the rainy season, the wastes decompose and provide nutrients for the tree.

Considerable quantities of silt are recovered in China, both from waterways and from the bottom of ponds used for fish culture. The silt is often composted with animal manures, green manure plants or aquatic weeds, and a little paddy straw. This is done under essentially anaerobic conditions in circular or square pits close to the fields. As the silt makes up some 80 percent of the input material, the final compost has a very low organic matter content, below 10 percent; although low in plant nutrients it is useful for improving the physical condition of soils. This method is described in detail in FAO (1978c).

4.4 COMPOSTING OF BRUSHWOOD, COFFEE PULP AND SEAWEED

It has been shown that composting a mixture of organic wastes proceeds better than handling a single material. The mixture supplies a wider range of the constituents listed in Table 3 plus appropriate enzymes, and probably supports a wider range of micro-organisms and small soil animals; it is possible that with a single material some species may be unattracted.

Nevertheless there are situations where single materials or types of material are present with little chance of mixing with substantial amounts of other wastes. Particular examples are brushwood and forest litter obtained from forest thinning, coffee pulp from coffee estates and factories, and seaweed from the seashore. If produced properly, compost made from these single wastes can make a good contribution.

4.4.1 Brushwood Composting

This technique was developed by Jean Pain working in a hot dry forested area in the south of France. He employed the material obtained during the thinning-out of the forests and the construction of fire-breaks: the broad avenues which reduce fire risk in such woodlands.

The material that can be used is essentially forest litter and undergrowth, plants and herbs, young tree seedlings and tree thinnings in the quantities collected. Where possible a high proportion of the material should have a twig diameter less than 8 mm and such material should be more fresh and green so that it will decompose fairly rapidly. Thicker branches are cut into thin slivers with machetes, pangas or a powered shredder.

The brushwood first requires soaking with water. This can be achieved by spreading it in a thin layer about 150 mm deep on the ground during rainy weather, turning the wastes over occasionally. Alternatively, water can be added by a watering can or spray. Pain found that the most suitable way is to pile the brushwood into a barrel, compact it by treading, keep it pressed down by a heavy stone and then fill the barrel with water; it is then left for up to 3 days. One cubic metre of brushwood can absorb up to 700 litres of water. The waste is then removed, drained and piled into a heap and compacted. The process is repeated until at least 4 m^3 of material have been gathered together; this is the minimum volume to ensure satisfactory evolution of heat and subsequent breakdown. Pain estimates that this quantity of waste takes about 3 labour-days to gather up and soak with water; it should yield nearly 2 tonnes of finished compost.

About 21 days after piling up the brushwood, the material has softened, warmed up slightly and the heap has sunk. The compost heap is now built. Using a fork with prongs the brushwood is dragged off the compacted pile, opened out and put to one side. The material has usually turned brown and smells sour indicating that decomposition has commenced, probably under slightly anaerobic conditions.

The opened-out material is then rebuilt into a heap of triangular cross-section measuring 2.2 m at the base and 1.6 m in height. The length depends upon the amount of material available. The wastes are forked onto the heap layer by layer without pressing them down. Next an insulating cover is built over the heap in a manner similar to that used in the Chinese high temperature stack described earlier. The cover is a 20 mm thick layer of soil, sand, leafmould or old compost. Finally the heap is covered with big leafy branches to protect it from rain, snow, sun or wind. Within a few days composting is well under way and temperatures up to 75 $^{\circ}C$ in the core have been reported. The construction of a brushwood heap is illustrated in Figure 26. Materials like brushwood break down fairly slowly and the oxygen requirement can largely be met by the rate at which air can diffuse into the mass from the atmosphere outside. Hence special ventilating passages are not needed, unlike the situation when composting putrescible agricultural wastes by the Indore or Chinese stack methods. However, the brushwood heap is very well insulated so that most of the heat which is generated is held in the core, and high internal temperatures can be reached.

After 3 months in the compost heap, some 4 months after starting the whole process, the compost is ready for use. It is not completely broken down and mature and probably has a slightly high C/N ratio. Hence it is used as soil cover and not for seedling growth. If the compost is required in a mature form the heap is turned, repiled and left for a further 6 months. Where the compost heap is erected from very woody material obtained from whole shrubs or young trees, it will probably take 1-2 years before it can be used.

In practice, the immature compost is applied on top of the soil, around the crop at the seedling stage, in a layer 70 mm thick; it is then covered with a layer 100 mm thick of litter from the forest. Pine needles have proved very effective for this purpose even though they are normally considered as virtually uncompostable. Shredded tree bark is another form of litter. Such litter reduces considerably the evaporation of water from the soil and enables the crop to be grown, even in arid areas, without

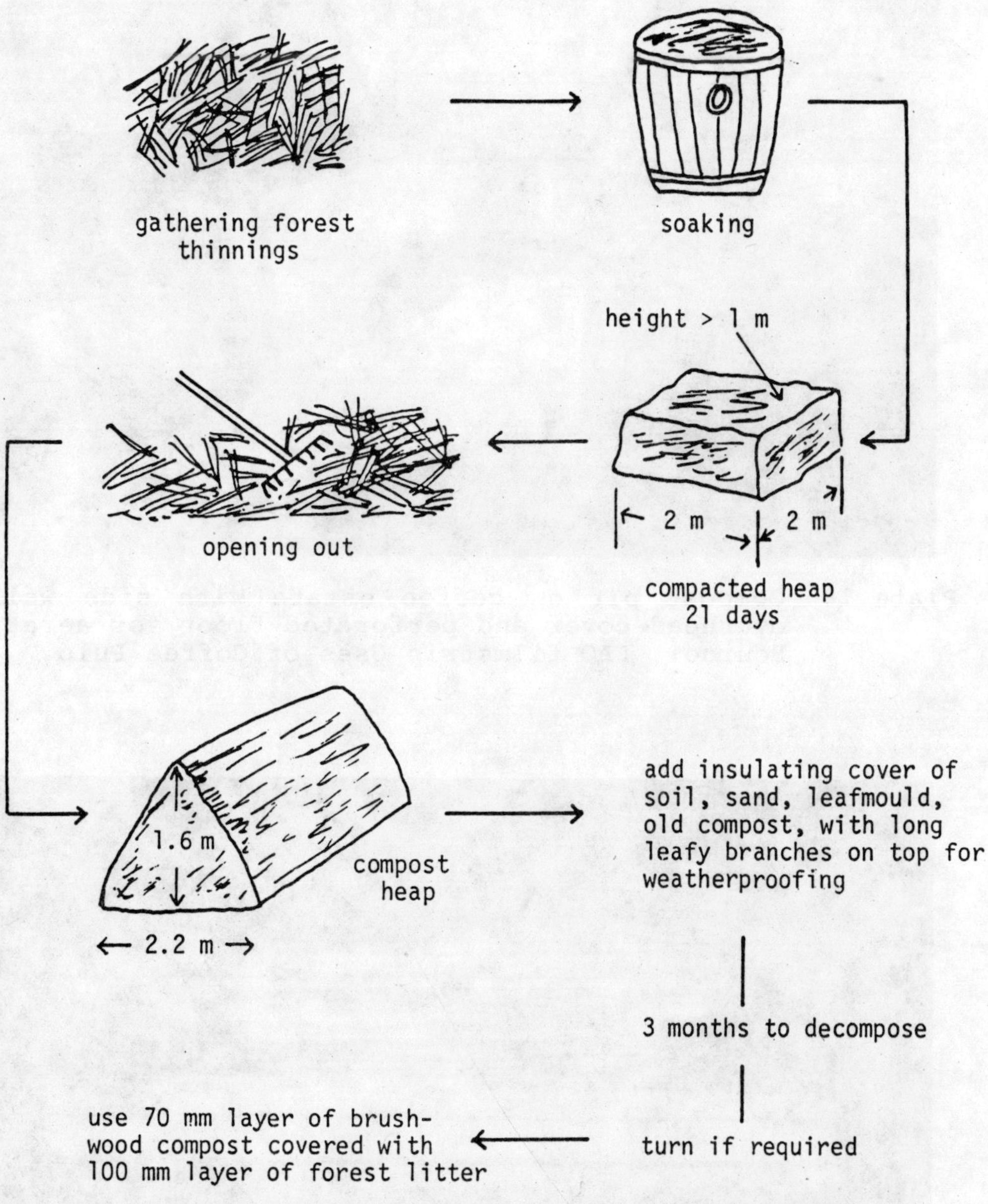

Figure 26 Brushwood composting system

watering subsequent to that at seed sowing. When pine needles or bark pieces are used as cover, they are removed at the end of the season and used in a subsequent compost heap. Where leaves, straw or grass are used, they can often be hoed into the soil at the end of the season.

The brushwood composting scheme has good potential as an alternative to the 'slash-and-burn' form of shifting cultivation. By composting the forest litter instead of burning it, valuable organic matter can be returned to the soil together with plant nutrients.

Plate 18 Compost pit on coffee estate with side walls, overhead cover and perforated floor for aeration
Source: FAO filmstrip Uses of Coffee Pulp

Plate 19 Details of bamboo floor of coffee pulp composting pit. Narrow gaps between the bamboo poles allow the pulp to drain and air to enter underneath the composting mass
Source: FAO filmstrip Uses of Coffee Pulp

4.4.2 Coffee Pulp Composting

Considerable quantities of coffee pulp accumulate on plantations or in factories where berries are prepared for market. This is becoming a serious environmental problem as the large heaps of material ferment, give off odours, breed flies and often pollute water courses.

Coffee pulp is a good fertilizer as it is rich in organic matter, nitrogen and potassium. Some growers spread the fresh or partly decomposed pulp around their coffee plants; however, this can cause transport and spreading problems with the heavy wet pulp and lead to odour and plant growth problems. It is much better to compost the pulp so that it can be used more effectively.

The composting unit consists of a number of well constructed above-ground pits as shown in Plate 18. This is roofed with corrugated iron sheets supported on bamboo poles; the roof prevents rain affecting the composting mass. The floor is made from bamboo poles mounted on bricks as shown in Plate 19; gaps between the poles enable the pulp to drain and air to pass under the mass before rising through it by the 'chimney effect' as composting proceeds.

The pulp from the berry processing unit is drained as much as possible and then loaded into the composting pit to a height of about one metre. Vegetable and animal wastes should be mixed in if available. A little soil or compost from a previous heap should also be added.

The material soon warms up and starts to decompose. The process can be helped by turning the heap, either from one section of the pit to another or into an adjacent empty pit. Less work in lifting is involved if the pit is divided into 2 parts by a cross wall of removable bamboos. The material is then turned from one half into the other. Such turning need only be done once every 4 to 6 weeks; the product should be mature and ready for use in 4 to 6 months.

4.4.3 Seaweed Composting

The sea is very rich in plant life and some use has been made of these resources for many centuries. Seaweed is an algae, a larger relative of the algae micro-organisms found to a minor extent in compost heaps. They differ from the more complicated plants in having no true leaves, stems or roots; the larger species, however, have organs which enable them to attach themselves to rocks. Many varieties of seaweed occur and can be collected from rocks or after being swept ashore. Although seaweeds naturally have a high moisture content, they are a valuable source of organic matter and minerals, especially nitrogen and potassium. They also contain a wide range of trace elements and are a source of plant growth hormones.

When seaweed is swept ashore it decomposes rapidly into a very smelly, slimy mass in which small flies breed. Composting is a better alternative. The seaweed is best washed with fresh water if it is to be used for sticky clay soils; this is not so important with sandy soils. It is then spread in a layer about 300 mm thick for two days; it thereby loses about half its moisture content through drainage and evaporation. The material is then mixed with an equal quantity of cereal straws, stalks or bracken to form an open matrix into which air can diffuse; some vegetable wastes and animal manures can be added if available. With practice, the seaweed can form up to half the weight of the mixture. The material is then built into a stack not more than 2 m wide and 2 m high with vertical ventilating holes formed by stakes not more than 1 m apart, as in the Indore process.

After 4 weeks the stack needs turning to bring the brittle and dried weed at the outside edges into the central core; a little watering may be needed but ventilation holes are usually unnecessary. Within 3 months from heap construction the mixture has decomposed into a fine, crumbly product.

The anaerobic method of breaking down seaweed is to mix equal quantities of seaweed and soil and heap this up for many weeks. This is still a useful practice where extra organic material for blending is very scarce. Seaweed can also be stored under a thick coating of soil for composting later, although it will give off strong odours when the pile is opened.

4.5 COMPOSTING OF NIGHT SOIL AND REFUSE

With increasing populations and migration from the countryside to the cities as a result of rural unemployment, towns and cities are facing increasing problems in disposing of night soil, sewage and refuse. Several papers in a recent conference (Schelhaas 1982) detail the difficulties facing some municipal authorities in Africa, Asia and South America.

Refuse disposal may be carried out by crude dumping, sanitary landfill, manual or mechanized composting, or incineration. Crude dumping will persist for some years due to the shortage of finance for improved methods. In less developed countries, the refuse is mainly organic in nature and will eventually decompose, while most metal and other objects are effectively scavenged and recycled back into the economy. Nevertheless, such dumping can lead to odour, flies, vermin and water pollution problems.

Flintoff (1976) suggests that for a low rate of waste generation, 400 grammes/person/day, manual composting costs twice, and incineration costs 15 times as much as sanitary landfill. As suitable sites for landfill are used up, composting will become more widely practised. Successful municipal composting depends upon several conditions:

i. the suitability of the wastes. Table 6 shows that refuse from tropical areas is far more suitable for composting than is that from Europe; it has a much higher vegetable/putrescible fraction and less uncompostable material;

ii. a market for the compost product within a reasonable transport distance, possibly 25 km, of the town or city;

iii. the support of the agricultural authorities, particularly the Ministry of Agriculture, so that local extension workers can recommend the use of the compost;

iv. a price for the product which is acceptable to most farmers;

v. a net disposal cost for the refuse (compost plant costs minus income from compost sales) which can be afforded by the local authority.

The production and use of municipal compost can also be justified for bringing into cultivation quickly marginal lands which are very short of organic matter. For instance, much of the compost produced in Holland in the nineteen-thirties and forties by the VAM Company was used to reclaim heath land for agriculture.

In some Chinese cities, such as Shanghai, much municipal waste is transported out to the countryside, often by boat, for composting by the communes rather than in large centralized city installations. However, with increasing quantities to be handled this practice is coming under pressure.

As mentioned in Section 3.2.9, in two cities in West Africa municipal refuse is used direct by local market gardens for vegetable growing, apparently with success. The refuse is of very high vegetable/putrescible content and has already been in the streets for several days prior to collection; some degree of composting has probably started already. Nevertheless such a practice cannot be recommended. Even piling the wastes for a few days aeration on a false floor, as shown in Plate 19, until the peak temperature is reached would significantly reduce any pathogen content.

The disposal or recycling of night soil requires very careful control because it frequently contains pathogens and the ova of intestinal worms. The practice of using it untreated on crops for human consumption is strongly to be discouraged, both in the interests of agricultural workers and of those consuming the food. Eventually, with rising standards of living, night soil will be far more widely used for biogas production. About 40 000 biogas units are in use in India and nearly one million in China. Such units greatly decrease the pathogen content of night soil and manures; however, worm ova often sink to the bottom of the digester and escape complete destruction. In China much night soil is transported out by boat or closed cart to the rural communes for composting; but some is composted by the municipal authorities for use within the town or city.

4.5.1 Simple Composting Techniques

In India in the nineteen-thirties, it was soon established that the Indore process was not the best method for handling large quantities of refuse and night soil in towns, due to the relatively high labour requirement and inadequate protection from rain, sun and wind.

Work under Acharya (1950) showed that the Bangalore method was more suitable. In this technique pits were filled with refuse and night soil and the material not turned during the composting period of 4 to 6 months. The pits were approximately 7 m long, 2 m wide and 1 m deep, the size being such that one or two pits could be completely filled per day; where possible the pits were lined with bricks or concrete. The pits were filled by putting in layers of refuse 200 mm deep which were slightly higher at the edges, then filling the central depression with night soil. The ratio used was approximately 150-300 litres of night soil per cubic metre of refuse. The layering continued until the pit was full when a final cover of refuse was added. The mass warmed up quickly to about 60°C and the initial aerobic decomposition was quickly followed by anaerobic conditions. As the material sank with early decomposition further layers were added. A 50 mm thick layer of soil was often put on top to help prevent fly breeding. After 4 to 6 months the product was dug out of the trench and screened to remove any large or uncompostable pieces. The Bangalore method appears to have been reasonably successful. It could not be employed where the water table was high, causing the pit to become flooded. It caused some problems with odour and fly breeding and a high degree of decomposition was not always achieved.

Gotaas (1956) gives a very detailed account of the Indore and Bangalore processes, together with other methods by Scharff in Malaysia, Wilson in East Africa and van Vuren in South Africa. The latter successfully used pits constructed above ground, with aeration floors, as shown in Plate 20.

On a village scale, the Chinese high temperature stack method described in Section 4.3.2, with ventilation channels and outer mud cover, is proving very effective at night soil and refuse composting. About a third of the material can be night soil, another third refuse, and the remainder should be strawy, stalky agricultural wastes if possible. The latter are helpful, both in holding open the structure of the mass to allow

Plate 20 Pits constructed above ground for refuse-night soil composting. Hammer mill for crushing and sieving facilities under cover at rear
Source: Gotaas (1956) and van Vuren (1949)

air movement and in absorbing the moisture from the night soil and that released by the refuse on breakdown. As shown in Table 6, the refuse in tropical countries normally contains mainly vegetable or putrescible wastes with little paper or cardboard to absorb moisture. The ventilation passages are important and can be made using bamboo poles, wood stakes or maize stems. The insulating cover is necessary to hold in heat, prevent fly breeding and absorb smells; it can be made of silt, soil, old compost or cereal straws such as paddy, or grasses. Straws and grasses should be raked off prior to turning and then used to cover the new heap or used as compostable material. In very cold and wet weather the leafy branches used in brushwood composting, Section 4.4.1, will provide further insulation. A good degree of aeration and insulation should enable good compost to be made hygienically from night soil and refuse in almost any weather; a minimum heap size of 3 tonnes, 6 m^3, is advisable under very bad weather conditions.

In a small town conditions are slightly different compared with a village. There is less access to strawy, stalky wastes for incorporation into the heap and to materials for providing the insulating cover. In addition much of the product compost will not be for use within the town but for transport and sale in the surrounding countryside. There will need to be close control on fly breeding and odour generation. Again the high temperature stack method is suitable. The refuse content will need to be about 70-80 percent by weight and the night soil 20-30 percent. Uncompostable items such as glass, metal and plastic should be removed from the refuse which is then spread out in a layer about 300 mm thick, the appropriate amount of night soil added on top and the two components mixed together. The mixture is next assembled into the above-ground stack in layers and the bamboo poles for aeration, or their equivalent, set in place; the stack is then completed. As an alternative to pre-mixing, the refuse can be laid out directly in the stack in layers 150-300 mm thick with the edges slightly raised by about 50 mm. Into the central depression is then poured the night soil; the moisture from this will gradually move to the heap edges. Covering the heap with an insulating layer is very important. For the first few heaps local soil, sand, coal ash or wood ash should be used but these may soon be used up. Thereafter the major source of cover material will be compost from an earlier heap; hence there may be no net production from such a compost site for some six months until an adequate quantity of cover material has been accumulated.

When using this composting technique in towns, it is advisable to complete a stack, or section of a long windrow, each day so that partially built heaps are not left exposed for fly breeding. Removal of the bamboo poles, heap turning and monitoring of progress should be done as mentioned earlier. In a stack containing only refuse and night soil, without any stalky wastes, air movement is more difficult. Hence it is likely that conditions in the heap will become slightly anaerobic in the early days; this should gradually improve with time. As long as the insulating cover has been made properly few smell problems should arise.

When the compost in a heap is judged to be mature and ready for use or sale, the heap is dismantled and the cover material saved if needed. The compost is then screened to remove large pieces; if these are of undecomposed organic material they can be used in another compost heap. However, if the reject material is mainly uncompostable items it is sent to the local tip or landfill.

4.5.2 Partly Mechanized Techniques

Partly mechanized techniques can be employed when towns can allocate more finances to waste disposal.

Many towns will operate a number of agricultural tractors towing trailers of up to 6 m^3 capacity; hydraulic tipping gear on the trailer is powered by the tractor engine. With a little practice the combination can be used to reduce a lot of the manual labour needed in transporting material to the compost stack. However, there will be less opportunity for removal of uncompostable items, apart from scavenging before refuse collection; hence such rejects will need to be removed by screening from the final product.

Actual construction and turning of the stack can be done manually or with a hydraulically operated bucket mounted on the front of a tractor. Night soil can be applied by hose from a tanker vehicle or trailer; the night soil may even have been pre-treated in a biogas digester and be available as biogas sludge. Any water required would probably be available from a pressurized piped supply.

Following composting the product can be screened by a power-driven rotary screen or trommel inclined at a small angle to the horizontal. Such screens have a good efficiency of separation as the wastes are tumbled for several minutes with the particles separating from each other. The screened size is likely to be in the range of 10 to 25 mm but a screen may be fitted with more than one size of mesh in order to achieve different grades of product. However, for efficient screening the compost must be fairly dry, below 40 percent moisture content on a fresh weight basis; with moister material the meshes quickly clog.

This whole subject of refuse composting at a more mechanized level is covered in detail by Flintoff (1976). He shows that the next stage of technical complexity for units handling about 50 tonnes/day involves equipment for refuse shredding, separation by screens, magnetic metal separators and ballistic separators. However, he strongly advocates the setting up of a manual pilot project first. Large-scale mechanized composting is considered further in Section 4.7.

In order to increase the rate of breakdown in windrow heaps and to reduce or remove the need for turning, some systems provide a forced air supply from a powered fan via perforated pipes mounted beneath the heap. This approach to providing the optimum air supply, given in Table 4, is justified where the area for windrows is inadequate for the throughput of refuse and faster decomposition is necessary. Care must be taken to avoid over-aeration of the heap which in hot climates will soon dry out the material and cause breakdown to cease.

4.5.3 Composting of Dewatered Sewage Sludge

In the treatment of waterborne sewage a sludge is obtained on settlement of the liquor. This sewage sludge is often partly dewatered by filtration. A number of forced aeration processes have been developed in the USA and France and are being used for the composting of dewatered sewage sludge with wood chips or sawdust as bulking agents. The Beltsville aerated pile process uses sewage sludge of about 78 percent moisture content with wood chips in the volume ratio of 1 to 2. The mixture is placed in heaps on top of perforated pipes and covered with a layer of finished compost; air is sucked through the pipes by a fan. The heap is composted for 4 weeks, then removed and stored for 4 weeks. The wood chips are then removed from the compost using screens. Use of the Beltsville process for the composting of night soil is described by Shuval *et al*. (1978).

Finstein *et al*. (1983) compare two 36 tonne heaps of sewage sludge and woodchips, one with air being sucked through the heap, the other with air being blown. Their recommendation is that air should be blown through the heap with the blower being switched on and off to maintain the temperature in the heap at 60°C. de Bertoldi *et al*. (1982) composted a mixture of sewage sludge with the organic fraction of urban refuse in three 2 tonne heaps that were identical mixtures but had different methods of supplying air. The refuse and sewage sludge were mixed in the proportion of 60 and 40 percent by weight respectively. Heap 1 was turned twice a week to provide adequate air. In heap 2 air was sucked through the heap with the fan working for 40 seconds every 13 minutes. For heap 3 a fan blew air through the heap for 40 seconds every 13 minutes when the temperature was below 55°C; the fan operated continuously when the temperature was above 55°C. The experiment showed that the method in which air was blown through the heaps gave a faster rate of breakdown and produced a compost product with a lower moisture content and a higher degree of maturity.

4.6 EQUIPMENT FOR MANUAL COMPOSTING

Equipment should be low cost and simple. Ideally it should improve labour efficiency, decrease monotony and be usable for other farming or local activities.

4.6.1 Transport of Bulky Wastes

The normal transport systems of the area will usually be adequate. Stretchers of the type shown in Figure 27 are cheap and effective for the collection of vegetable wastes and loading them into bullock carts for transport. The capacity of a bullock cart for carrying bulky materials can be increased by building a framework of light poles on it and covering the sides with any locally available matting or sacking. A modified cart for use with a single bullock has been developed in South India for transport of bagasse from sugar cane crushers to drying yards and back to the cane juice boilers. Its capacity is 2 m^3 and it can carry loads of up to 500 kg.

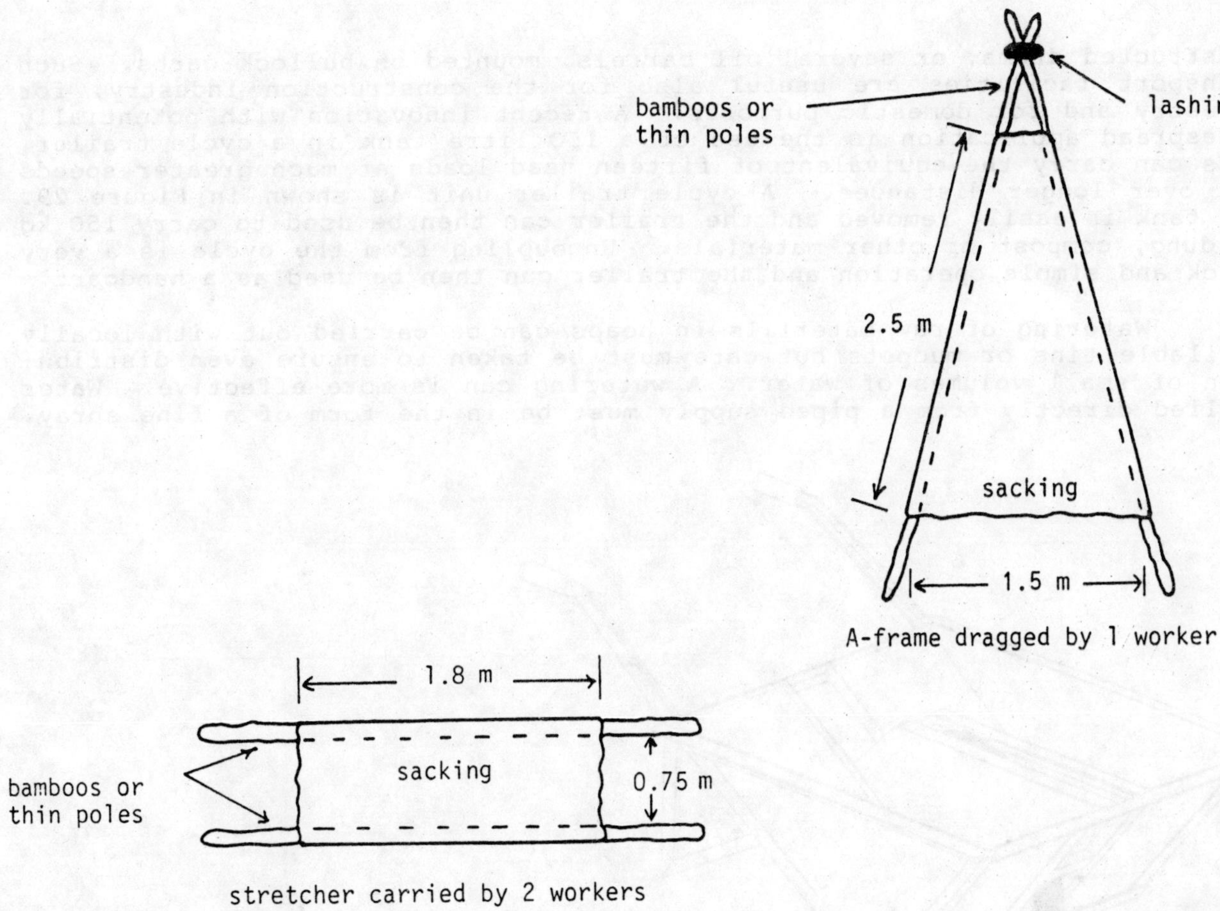

Figure 27 Stretchers for carrying wastes

The body of the cart tips easily on manual operation of a lever and this helps quick and easy unloading of the bagasse. The cart is equally effective for transporting bulky raw materials and the compost product.

4.6.2 Transport of Dense Materials

In many areas dung is carried in small baskets on the head with individual loads of approximately 5-7 kg. Such work is unpleasant for the operator. Simple handcarts or wheelbarrows can markedly increase the efficiency of the operation. The small farm transport vehicle has been designed on the engineering principles of the Chinese wheelbarrow (Figure 28). It can be made from wood or steel in very modestly equipped workshops and can be fitted with a wide variety of wheel types and body cladding according to local need. It has been used on a regular basis in India to carry loads in excess of 100 kg over distances of up to 1.5 km in a one-man operation. The equipment is suitable not only for other farming activities but also for the local building and road construction trade, as well as for many other transport operations involved in marketing.

4.6.3 Handling of Water

This is a very important part of the composting operation as an adequate supply of moisture evenly distributed throughout the heap is a major factor in achieving a high quality product. It is most unlikely that a piped water supply installed for composting will be justifiable economically; hence siting of the composting area should be considered in relation to the available water supply. Water can be transported in specially constructed drums, or several oil barrels, mounted on bullock carts. Such transport facilities are useful also for the construction industry, for forestry and for domestic purposes. A recent innovation with potentially widespread application is the use of a 150 litre tank in a cycle trailer. This can carry the equivalent of fifteen head loads at much greater speeds and over longer distances. A cycle trailer unit is shown in Figure 29. The tank is easily removed and the trailer can then be used to carry 150 kg of dung, compost or other materials. Uncoupling from the cycle is a very quick and simple operation and the trailer can then be used as a handcart.

Watering of raw materials in heaps can be carried out with locally available tins or mudpots but care must be taken to ensure even distribution of small volumes of water. A watering can is more effective. Water applied directly from a piped supply must be in the form of a fine spray.

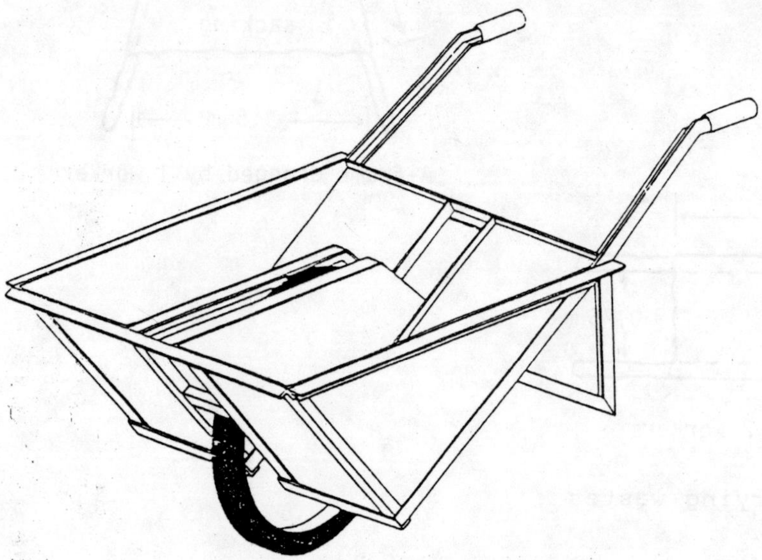

Figure 28 Small farm transport vehicle (I.T. Transport Ltd)

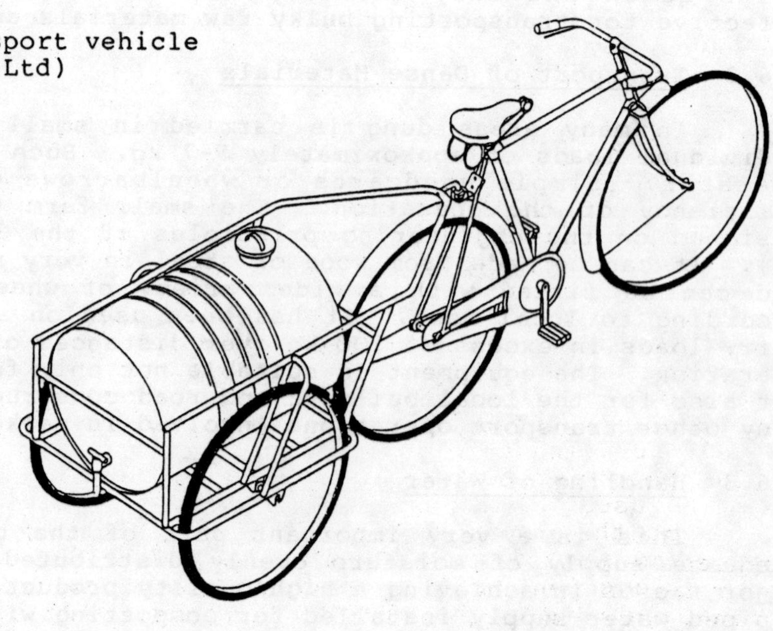

Figure 29 Cycle trailer unit (I.T. Transport Ltd)

4.6.4 Preparation of Raw Materials

A reduction in size of the material to the standards suggested in Section 2.4.3 is ideally achieved using a purpose-built cutter or shredder. Such machines may be expensive to buy and normally need a power supply. Hand operated cutters tend to have low output rates. While in some places the cutting/shredding operation has been coupled to pedal power, there is as yet no commercially available pedal powered machine. Pretreatment using a lime mortar mill powered by one or more bullocks (Plate 8) can be a very effective means of crushing stiff stemmed materials or urine-earth from cattle sheds; however, it is unlikely to be economic unless the mill has been installed primarily for mortar processing. Chaff cutters can be used for reduction of particle size but where no other facility is available, then stalks and straws can be chopped up using machetes, sickles or similar traditional hand tools. Such tools can be quite effective and have acceptable output rates when used to chop small bundles of stalks.

4.6.5 Making and Turning Heaps

Material handling can be done with locally available baskets, spades and rakes. Light poles up to 3 m long and 100 mm in diameter will be necessary to establish and maintain air vents. If poles are not cheaply available then bundles of maize or sorghum stalks can be used.

4.6.6 Sieving Product Compost

From a reasonably mature heap compost contains a wide range of particle sizes from fine grains to partly decomposed twigs and uncompostable fragments from refuse. For some uses compost made from agricultural wastes may need no sieving, just the removal of large twigs by hand as the material is spread as a surface mulch. For use when seed sowing the compost can be sieved using a screen mounted at a slight angle to the vertical. The form of sieve used for screening sand in the construction industry, but covered with weld-mesh or chicken wire, is suitable for compost; the hole size should be about 25 mm. With composts prepared from refuse, a smaller hole size of 10 mm should remove most of the unwanted rejects of glass, metal and plastic.

On a very small scale the open-work baskets used for winnowing crops such as paddy can be employed for sieving compost.

4.7 LARGE-SCALE MECHANIZED COMPOSTING

Modern large-scale urban composting systems suitable for towns and cities usually comprise refuse storage facilities, feed preparation equipment, the breakdown stage and final product upgrading.

Refuse collection vehicles discharge their contents into pits or onto flat concreted areas from where it is conveyed by moving floors, overhead grabs or front end loaders or mechanical buckets. The material is then reduced in size, unwanted and salvageable material removed and the moisture content adjusted. Size reduction can be carried out in wet pulverizers which consist of an inclined drum about 3 m diameter and 10 m long, rotating at 3 to 10 revolutions per minute with a throughput of about 10 tonnes per hour. Alternatively, dry pulverizers such as hammer mills may be used. The power requirements of pulverizers depend mainly upon the size of the outlet material; for reduction to 50 mm about 8 kWh/tonne is required whilst reduction to 12.5 mm requires about 20 kWh/tonne. After size reduction, magnetic separators remove ferrous metals; sheet plastics, rags and oversized material are removed by screening machines. The moisture content is adjusted, if necessary, by the addition of water or sewage sludge before the wastes enter the breakdown stage.

The breakdown stage is carried out in simple windrow heaps, windrow heaps with forced aeration, or in more sophisticated mechanized units. The latter vary from automated windrow systems with continuous automatic turning equipment, to totally enclosed vessels which achieve extensive breakdown of the material within a few days.

In automated windrow systems the material is usually placed in troughs or between walls. The base of the composting area has the necessary pipework for the blowing or sucking of air. Agitation is provided by a turning device which moves along the walls on rails or wheels. Various designs of turning device have been used. Some consist of a large paddle wheel 4 m in diameter which picks up the composting material and places it on a central conveyor which transfers the material to form a new heap. Other designs have screw conveyors which gradually move the material towards the final processing stage. The residence time in these heaps varies from 4 to 12 days, depending upon the design.

Rotating drum units such as the widely used Dano system consist of a cylinder up to 4 m diameter and 40 m long, inclined slightly to the horizontal. Refuse is fed to the higher end of the drum which is rotated continuously at about 1 to 2 revolutions per minute. Various sizes of screen mesh may be built into the sides or the end of the drum to separate the material according to size. The larger size rejects may be separately disposed of or returned to the inlet of the drum. The residence time in the drums is usually 2 to 3 days after which the material is placed in windrow heaps for several weeks. These heaps may be turned 2 or 3 times.

Vertical tower systems may be circular or rectangular in cross section. The simplest type of circular tower has one floor and is filled each day full of refuse. A number of such towers are provided to give the required residence time of 4 to 5 days. Air is usually supplied by blowing through holes in the floor. Agitation may be provided either by vertical screw devices or by horizontal rotating arms.

In multi-floored towers, of typical size 8 m diameter by 12 m high, the refuse is fed in at the top in approximately 50 tonne amounts and slowly moves downwards. The floors may be in sections which are hinged to open periodically and drop the material to the floor below. Some designs have holes in the floors with rotating arms that squeeze the material through the holes to the floor below. A more complicated design has trap doors in each floor which open to allow rotating arms to push the material through to the floor below. Air is provided either through the floor or through the rotating arms. Water sprays may be placed above each floor to adjust the moisture content.

A single tower may contain up to 200 tonnes of material which may release substantial amounts of heat, causing the temperature to rise above 60°C. Excess air is often provided to cool the material by evaporating moisture from the wastes into the air stream. In order to prevent the material from becoming too dry water sprays are often used. The flow of air may be controlled separately at each floor. The extent of instrumentation and automation provided in tower systems varies according to the design. Residence times in the towers are in the range 4 to 20 days, 8 days being most common. Some designs provide for windrow heaps after leaving the towers while others have a simple storage area of 2 to 3 weeks capacity.

In multi-floored rectangular shaped compost houses the refuse is fed into one end of the top floor and is moved along the floor by a conveyor or screw device. The material then drops to the floor below and moves in the opposite direction. Air is provided through the floors. An alternative design uses sections of the floor which turn over and drop the material to the floor below; aeration is achieved as the material falls from floor to floor. Residence times are typically 4 to 8 days, usually followed by a maturing time in heaps.

Large-scale mechanized systems composting refuse are likely to have some form of further treatment of the compost product before use. In many cases, screens with 25 mm holes are used to remove any remaining plastic and textiles. Large glass fragments may be removed together with any other inert particles. The material less than 25 mm may be further screened to produce a very fine product below 8 mm size. Another form of final treatment is to pulverize the compost in a hammer mill. In this way glass fragments, often a cause of complaint, are reduced to sand-like particles.

Since the introduction of large-scale composting systems in the nineteen-fifties some 30 different processing schemes have been introduced for composting city and town wastes, with varying success. Equipment for feed preparation and compost product finishing are similar in many of these processes. The breakdown stage, however, has varied widely, being carried out in different designs of pits, vessels, towers and drums. Very recently large-scale mechanized systems with capacities in the range of 200 to 500 tonnes of refuse input each day have been installed in various countries. Enquiries have been made regarding 750 to 1000 tonne per day factories. These large mechanized units are designed and installed by specialist contracting companies; they produce very large quantities of compost but are capital intensive and require skilled technical support for their operation and maintenance.

In recent years, a significant number of mechanized units have been erected in France and the USA for the composting of dewatered sewage sludge with a bulking agent such as sawdust or wood chips. The units employ closed vessels instead of the large windrows of the Beltsville technique. The vessels are either circular or rectangular, without intermediate floors; aeration is supplied from the bottom and specially designed screw conveyors remove the finished product once it has reached the bottom of the vessel.

4.8 CONSIDERATIONS WHEN CHOOSING A PROCESS

In Chapter 2 the principles controlling the composting process were described. The present chapter has shown how these principles are put into practice. Although the householder, farmer or municipal authority may not have available an ideal range of organic wastes for composting, much can be done by taking care with building and looking after the heap. The importance of good aeration arrangements and a suitable moisture content have been emphasized, together with the need for heat insulation; the latter is particularly important when composting small quantities of material.

The choice of a particular process depends on several factors:

i. the quantity and nature of the organic wastes;

ii. geographical considerations such as the climate, height of the water table and the need to reclaim eroded hill slopes;

iii. the skills and availability of the necessary work force;

iv. the finance available for both capital and operating costs;

v. the major use for which the compost product is required, such as for farms, market gardens, nurseries or land reclamation.

A current major consideration is that being experienced on arid hill slopes in East Africa where deforestation and overgrazing have led to considerable loss of topsoil and vegetation has virtually gone. Here the Mazibuko or 'fertility' trench system can help agriculture start again. However, it is probably beyond the capabilities of one smallholder and needs a cooperative programme with several farmers or a whole village both to dig the trenches and to supply the organic wastes. In practice the trenches are filled in sections as wastes become available, the section ends being held in with corrugated iron sheets. The growing of maize in the locality is important as this provides a large volume of stemmy, woody material which breaks down slowly in the trench. It may take 5-10 years to revegetate a badly eroded hillside but once done the effects should be long lasting.

In less seriously affected places where wastes from agriculture, trees and roadside vegetation are reasonably plentiful, the Indore process,in pits or stacks according to the time of year, is still very suitable. Putting the wastes in layers is important to avoid large areas of single materials, especially ones which are not easy to aerate. The process is much improved by covering the stack or top of the pit with a layer of soil, banana leaves or woven palm leaves to improve heat retention and reduce drying out. As mentioned in Section 4.3.1, turning of the heap can be adjusted to the labour availability at the time of the year.

The composting of refuse does not usually lead to major environmental problems. Hence, where refuse is among the wastes it can often be handled by the Indore process, particularly if it is mixed with agricultural wastes.

However, the treatment of night soil introduces definite problems due to its likely content of pathogens and great attraction for flies and their breeding. Therefore, where night soil has to be treated in any appreciable quantity, the Chinese high temperature stack with an adequate cover is strongly recommended. The composting area needs to be kept tidy and spillages of night soil avoided.

In a village it is wise to keep to simple and easily understood techniques of heap construction until some degree of expertise has been acquired.

In a town, where mainly refuse and night soil are to be composted, the site for operations needs to be well chosen. It is best placed about one kilometre outside the town to reduce complaints in case odours are generated accidentally. For a fairly large town, two sites on opposite sides will minimize transport hauls. The site needs to be reasonably level with a firm surface which is well drained. When operations are partly mechanized, with tractors being involved in turning operations on compost heaps, the surface can quickly be churned up in wet weather. Good drainage is important to prevent the lower parts of heaps from becoming waterlogged in the rainy season.

In addition to the area laid out in compost heaps, a storage area for product compost is needed; this is because farmers tend to use compost only at certain seasons. In this area the product can be piled to a height of several metres.

The site is best surrounded by a hedge or row of trees to serve as a wind break to prevent minor smells being carried away and to reduce the blowing of paper, dust and litter around the area.

When some amount of capital expenditure can be afforded, above-ground or sunken pits can be erected on a more permanent basis, with or without overhead protection; examples are shown in Plate 18 for composting coffee pulp and Plate 20 for refuse/night soil mixtures. Even in these cases, it is wise to start operations using a simple stack technique until experience at composting has been gained.

With cities requiring composting units of high capacity, transport hauls will probably dictate siting the factory close to the suburbs. Great care needs to be taken to reduce the chances of complaints arising due to smells, flies or litter. This may well require the use of enclosed fully mechanized units instead of windrows. The land requirements for a mechanized process depend on the type of plant and the time allowed for maturing and storing the product. A highly mechanized tower-type unit handling 500 tonnes per day of refuse, with product storage capacity of 14 days, could be constructed on an area of 1-2 hectares. A windrow system of similar capacity could require 10-15 ha, depending on the residence time in the windrows. The cost and availability of land may therefore be a factor in process selection.

Highly mechanized processes require the attention of skilled maintenance engineers and workshop facilities. Simpler processes with a larger requirement for manual labour need less skilled operators. Thus, to some extent, the skills of the available work force affect the choice of process.

The quality of the compost product which is required by the customers will influence the extent of the finishing operations. Compost for upgrading low quality land need only be very roughly screened. Compost prepared from municipal wastes needs to be more finely screened when used for horticulture than for agricultural purposes. The acceptability of visible fragments of glass and plastics in the product will vary from customer to customer.

Usually the most important consideration is that of capital and operating costs. These will vary considerably depending on the type of process used and the country of operation. Costs of land, power and labour, and the financial returns to be made from sales of the compost, differ considerably. The present day capital cost of large-scale composting units, including civil engineering site works and all necessary ancillary equipment, is probably in the range of (sterling) £15 000 to £40 000 per tonne per day of refuse input.[1/] Windrowing plants would be at the lower end of the range, highly mechanized tower units at the upper. Irrespective of the simplicity or complexity of the breakdown stage, most modern processes include equipment for shredding and separating the incoming refuse and for screening the final product.

The amount of labour required in composting depends on the degree of mechanization employed. The Indore process was originally estimated to require 3 labour-days to prepare and spread 1 tonne of compost product. With improvements in technique this was later reduced to 2 labour-days per tonne. Using the Bangalore process of composting night soil and refuse in pits without turning, 1 labour-day per tonne was adequate; however, this slow process took more land and the digging of many pits. For the Chinese high-temperature stack method, with its need for placement and removal of bamboo poles and the addition of a mud cover, about 2 labour-days per tonne can be used as an estimate. Where some mechanical equipment can be used, such as tractors with front buckets for turning operations, then labour needs can be reduced considerably.

1/ 1986 prices. Dollar equivalent US$1.4 to £1.

The breakdown of organic materials during composting results in the loss of about 40 percent of the organic matter to the atmosphere as carbon dioxide and water. Thus the weight of compost product produced is significantly less than that of the initial wastes. There will be additional losses as rejects from any separation and screening processes, especially when refuse is being composted. As a very rough approximation, the compost produced from agricultural wastes will be about 40-50 percent of the weight of input material, that from refuse and night soil composting in towns and cities will be about 40 percent after reject materials have been removed by screening.

CHAPTER 5

USES OF COMPOST

5.1 GENERAL

The previous chapter describes a number of well-tried composting processes for the treatment of organic wastes from agricultural or municipal sources, or a mixture of the two. However, unlike landfill or incineration, these composting processes are not solely techniques of waste disposal; they are intended to prepare the wastes for return to the soil as part of the Cycle of Life (Figure 1). Via composting the wastes should be prepared for rapid assimilation by the soil with minimal disruption of soil processes and maximum benefit in maintaining, or indeed improving, soil fertility. Hence wise use of the compost is just as important as choosing the best methods for its production.

Accordingly, it is necessary to look at the various needs for, and methods of, improving soil fertility and seeing what part the use of compost can play in the programme. The composition and properties of compost on a biological, chemical and physical basis need to be examined. This helps in deciding how to use the compost for vegetable growing, tree planting and field crops.

The time taken for compost to be ready for use depends partly on the care taken in its preparation, in terms of aeration, moisture control and amount of turning, and partly on the use for which the compost is required. One may need rough compost for reclamation of very poor soils, a medium grade for general agricultural use and mulching, or a fine well-matured form for small seed sowing and for grass lawns. In the tropics composts can be ready in anything between 2 and 6 months depending on the above factors.

When compost is ready and is waiting in storage it should be carefully covered to protect it against heavy rain which will wash out its water-soluble nutrients, and against strong sun which will cause unwanted oxidation of the organic matter. The cover can be of fresh or dried banana leaves, woven palm leaves or a piece of sheet plastic.

Occasionally there are reasons for using compost material without delayed storage. In Western Samoa in the Pacific Ocean, compost heaps need to be used within 3 months of their construction; if left longer they become breeding sites for the rhinoceros beetle (<u>Oryctes rhinoceros</u> L.). In many parts of the tropics the termites or white ants can be a problem at certain times of the year; they can be attracted to compost heaps and build their nests in them. Hence, it is important to prevent termites gaining access to stored compost.

5.2 IMPROVING SOIL FERTILITY

Soil management should improve soil fertility so that crop yields can be increased and then maintained at these higher levels. A fertile soil is able to withstand erosion and meet the needs of the crop for moisture, air, nutrients, acidity and temperature. Fertility is aided by a wide range of farming practices, all of which help to maintain or increase the amount of organic matter in the soil.

5.2.1 <u>Erosion Control</u>

Erosion is the removal of fine particles of soil and organic matter from the land surface by fast moving wind or water. Less fertile coarse materials are left behind and the soil becomes shallow, low in plant nutrients and dries out quickly (Plate 21). In extreme cases the soil is either completely removed, exposing the underlying bare rocks, or a desert is formed with unstable shifting sand dunes. The material carried away is often deposited in waterways and dams, choking them and leading to reduced storage capacity and the danger of flash floods during heavy or continued rains. The first aim of tropical soil management is therefore to prevent erosion; this is achieved using the following principles:

i. reducing the destructive force of the wind and water. In the case of wind this can be achieved by growing shelter belts of trees and shrubs. If the species grown are selected for their own productivity of fodder, fruit and fuel, then no serious loss of output will result. The destructive force of water on croplands can be minimized by bunding and contour ploughing which prevent water from running at high speed down steep slopes and picking up large quantities of soil. In any system of contour ploughing it is important to make provision for controlled drainage to carry away excess water without causing erosion during particularly heavy rainstorms.

ii. protecting the surface of the soil by covering it with a mulch. This can be of organic residues (Plate 22) or artificial materials such as polythene. It is particularly important to absorb the energy of large raindrops and prevent them pounding the soil and displacing the fine particles. Plant leaves perform this function and hence the quick establishment of crop cover is an important means of protecting the soil on agricultural land. Tree cover is very important on non-crop lands.

iii. increasing the porosity of the soil so that more rain is absorbed and less runs off. Light pre-monsoon tillage can halve the run-off of the first rains. The addition of organic manures and composts will result in improved soil structure by binding soil particles into crumbs and increasing the number and sizes of pores which let water into the soil. The crumbs themselves do not break down easily on wetting and hence resist erosion. The Mazibuko or fertility trench system described in Section 4.3.3 is very effective at water absorption, the organic matter within it acting as a big sponge; the trapped water is then slowly released into the rooting zones of the planted crops.

Improved soil structure has many other benefits among which the prevention of soil capping is very important. The formation of a strong impervious cap or crust on the drying of sandy soils is a common occurrence which can seriously reduce aeration of the soil, crop emergence and the quick even establishment of crop cover. Breaking the crust by tillage may do as much harm to the crop as leaving the crust. Hence the prevention of crust formation is a very real benefit which results from the use of composts and organic manures.

5.2.2 Moisture

Moisture is of critical importance to crop growth and tropical soils must be managed to provide adequate levels of moisture to the crop under conditions which may vary in a few hours from drought to flood. Apart from the danger of erosion, flood conditions are likely to wash or leach large quantities of plant nutrients downwards below the reach of roots and to seriously reduce the availability of oxygen to roots. Very dry conditions cause the crop to wilt and also interfere with the uptake of nutrients from the soil.

Plate 21 The effects of water erosion: the topsoil has gone leaving infertile soil and bare rock
Source: FAO photo (F. Botts)

Plate 22 Mulching around established coffee plants with compost to reduce moisture loss and improve soil fertility
Source: FAO filmstrip Uses of Coffee Pulp

The best method of improving the moisture holding capacity of the soil is to add organic manures and composts, either as surface mulches or by incorporation into the top layer of the soil. The resulting improvement in structure will optimize soil aeration and enable the soil to hold more water before drainage starts or waterlogging takes place. Although at times of heavy rain the water will drain downwards beyond the crop root zone, the water-soluble nutrients will be bound to the organic matter and leaching losses will be minimized.

Use of organic manures, in conjunction with the contouring and drainage techniques referred to above, will lead to a marked lowering of the risk posed by too much or too little rain. It is important to realise that quite small improvements in the flood and drought resisting ability of the soil can make major differences to crop yield; they may well mean the difference between harvesting a reasonable crop and total crop failure. A rain-free period of 21 days will mean total crop loss on a soil which cannot support a crop, without permanent wilting, for more than 19 days. Crop recovery and harvest, albeit reduced, will be possible with an improvement to 23 days in the ability of the soil to resist permanent wilting.

5.2.3 Plant Nutrients

The nutrients required by crops in greatest quantities are nitrogen (N), phosphorus (P) and potassium (K). In addition many other minor nutrients or trace elements are required in smaller quantities; among these are calcium (Ca), sulphur (S), iron (Fe), boron (B), molybdenum (Mo) and zinc (Zn).

As the crop grows it removes some of each nutrient from the soil and on harvesting these nutrients are usually taken away from the field in grain, vegetables or crop residues such as straw and stalks. Nutrients which are water-soluble are also taken out of crop root zones by the downward movement of water; this is especially true of nitrogen in the form of the nitrate ion. Many of the required nutrients may be present in the soil but are in water-insoluble forms which are not usable by the crop. Soil nutrition must deal therefore with supplying nutrients, minimizing leaching losses and increasing the availability of these nutrients from the soil water solution to the crop.

The supply of nutrients is a very important aspect of soil fertility and there has been a tendency to assume that, as long as the crop's need for the major nutrients is met by mineral fertilizers, continued improved yields will ensue. Experience has shown that this is not the case. This is because mineral fertilizers do not supply all the nutrients needed by plants and because they have very little effect on other aspects of soil fertility such as structure, resistance to erosion and moisture retention.

Initially the use of mineral fertilizers and improved seeds will increase crop yields. This improvement inevitably requires greater quantities of trace elements not supplied by the mineral fertilizers and on poor soils trace element deficiencies, especially of iron and zinc, may soon develop. It is possible to augment trace nutrients, usually by spraying, but overdosing can easily occur and may be harmful. A more satisfactory approach is the use of compost which will normally contain all the trace elements required, in addition to useful quantities of nitrogen, phosphorus and potassium.

There are two difficulties about using compost as a source of plant nutrients. Firstly, if it is not fully matured the compost may have its own demand for nutrients as breakdown to maturity continues in the soil. In this situation the compost micro-organisms may rob nutrients, especially nitrogen, from the soil with harmful effects on crop growth. This danger is especially likely with compost made from raw materials which have a very high C/N ratio; hence care should be taken to ensure that such composts mature fully before incorporation into the soil.

The second difficulty is that, compared with mineral fertilizers, composts are dilute sources of N, P and K. It may not be possible to prepare enough compost to meet all the needs of the crop for these major nutrients. There is increasing evidence from many parts of the tropical world that there are definite benefits to be obtained from using compost and mineral fertilizers in conjunction. Thus, 30 kg of nitrogen from a mixture of compost and mineral fertilizer gives a better crop response than does 30 kg from compost alone or 30 kg from mineral fertilizer by itself. It is better to spread compost together with mineral fertilizer on all one's land than to use compost only on half of one's land and mineral fertilizer only on the other half. This is partly because mineral fertilizer is mainly water-soluble and is almost immediately available to crops, whereas the nutrients from compost are available more slowly over a longer period. In addition, the improvement in soil moisture characteristics resulting from compost incorporation increases the efficiency of use of mineral fertilizer. In a very dry soil the plant will not be able to take up nutrients and in a very wet soil the nutrients may be leached out.

Losses from leaching can be markedly reduced by increasing the soil organic matter content. Some nutrients in the water-soluble forms required by plants are readily leached from soil particles, whereas they are held effectively on the surface of organic matter in the soil. This does not seriously reduce their availability to plants but does prevent their loss by leaching. Reference to Figures 4 and 6 shows how losses of nutrients by leaching from the root zones can be made good by the recovery of the nutrients, or the mineralization of fresh nutrients from deeper soil zones, via the roots of trees or deep-rooting shrubs.

Most of the naturally occurring nutrients in the soil are held in forms which are not available to plants, hence crop growth is restricted by the rate at which these nutrients can be converted and dissolved into the soil water, or mineralized, in soluble forms which are available to crops. There is increasing evidence to show that the addition of organic matter to the soil, especially in the form of compost, increases the rate of mineralization of nitrogen, phosphorus and potassium. There is also increasing evidence that trace elements which are applied to correct deficiencies will be more readily available to crops and will exert a longer lasting effect if compost has also been given.

5.2.4 Soil Reaction (acidity/alkalinity or pH)

The reaction of a soil depends largely on the minerals from which it is composed and its history, especially whether or not it has been flooded. Every soil has a reaction or a pH value and the type of crops which can be grown on it depend largely on that value. A soil with a pH value below 7.0 is acid and the lower the pH value the greater the acidity. A pH value above 7.0 indicates an alkaline soil and the higher the figure the greater the alkalinity.

Soils which are either very acid or very alkaline will need special chemical treatments to bring them into the pH range for normal cropping. After soil reclamation it is important to maintain as high a level as possible of organic matter to prevent a recurrence of the original problem. This is especially important with irrigated soils where there is a danger that salts introduced to the soil in irrigation water will be deposited in the top soil if drainage is impeded due to poor soil structure. A range of crops suitable for soils of varying reaction is given in Table 13. For further information on this topic see FAO Soils Bulletin No. 31 and Irrigation and Drainage Paper No. 29 Rev. 1.

Table 13 SOIL REACTION (pH) RANGES OF VARIOUS CROPS

Crop	Optimum pH range (pH measured in water)
Cereals	
Maize	6.0 - 7.5
Millets (other than sorghum)	5.0 - 6.5
Rice	4.0 - 6.0
Sorghum	6.0 - 7.5
Wheat	6.0 - 7.5
Legumes & oilseeds	
Beans	6.0 - 7.0
Berseem	6.0 - 7.5
Grams	5.5 - 7.0
Groundnuts	5.3 - 6.5
Linseed	5.5 - 7.0
Mustard	5.5 - 7.0
Peas	5.5 - 6.0
Sesamum	5.8 - 8.0
Soybean	5.5 - 7.0
Sunflower	6.5 - 8.5
Vegetable Crops	
Carrots	5.7 - 6.5
Cauliflower	5.5 - 6.5
Cucumber	5.5 - 6.5
Onions	5.7 - 6.5
Potato	5.0 - 5.5
Tomato	5.0 - 6.0
Spinach	5.8 - 7.0
Plantation and other crops	
Cocoa	4.5 - 7.0
Coffee	6.0 - 6.5
Cotton	5.0 - 6.5
Jute	5.8 - 6.4
Oranges	6.0 - 8.4
Tea	4.0 - 6.0
Turmeric	5.5 - 6.5

5.2.5 Temperature

It is widely accepted in tropical agriculture that rapid and even seed germination has a very important bearing on achieving high yields, especially with short duration crops. One of the main causes of uneven germination and of low seed germination percentages is very high soil temperature. It is now believed that soil temperature is reduced by high levels of organic matter in the soil and as a result germination is improved. The use of compost will help in this respect. However, very immature compost should not be employed for this purpose; the organic acids produced initially during its continued breakdown in the soil, and the ammonia liberated later, will both inhibit the germination of small seeds.

5.3 COMPOST COMPOSITION AND PROPERTIES

5.3.1 Compost

The produce left in the compost heap (Plate 23) at the end of the breakdown stage consists of:

i. the more resistant parts of the original wastes, particularly the lignin or woody materials;

ii. intermediate products formed during the process;

iii. the bodies of dead micro-organisms;

iv. humus, the complex and stable material formed by complicated chemical interactions between materials i, ii and iii;

v. living micro-organisms and small soil animals;

vi. mineral matter (ash) brought in with the original organic wastes as soil on plants roots or road sweepings in refuse;

vii. water still present, mainly absorbed in the organic matter.

The amount of humus present depends largely on how long the product has been allowed to carry on into the maturing stage. Humus formation, or humification, is an extremely complicated process which is not completely understood. In very simple terms it is the combination of protein from the bodies of the dead micro-organisms with the lignin residues to form 'ligno-protein'.

Reasonably mature compost is a brown-black crumbly material (Plate 24). When made from agricultural wastes at least 80 percent will pass through a screen with holes 5 mm in diameter. A few small stalky and strawy remnants will be recognizable; otherwise the character of the original wastes will not be distinguishable. It will have an earthy smell.

If left to itself in the heap, the living micro-organisms will continue the humification process until complete maturity has been reached; at this point the C/N ratio in the compost will be close to 10/1. Thereafter the micro-organisms continue to feed on the humus, giving off carbon dioxide and wastes to the atmosphere and releasing the nutrients and other minerals contained in the humus as shown earlier in Figure 5. This process is termed mineralization; it takes place very slowly in temperate climates but far more quickly in tropical temperatures and especially when exposed to strong sunlight. Eventually all the organic matter will have gone, leaving only the mineral matter that it, and the original wastes, contained; however, some of the nitrogen will have been lost.

The same processes occur when the compost is mixed in with the soil. The compost not only provides food for its own micro-organisms but also for those in the soil with which it is mixed. Plant nutrients released become available for uptake by the roots of crops and trees. However, the rapid breakdown of organic matter under tropical conditions means that soils in these environments require regular inputs of compost, yearly if possible, so that the biological, chemical and physical contributions of the material to soil fertility can be maintained.

5.3.2 Biological Aspects

At the end of the breakdown process compost still carries an enormous population of micro-organisms, similar to the numbers shown in Table 2. The application of compost not only adds much microbial life but

Plate 23 Finished compost in a heap; it is black-brown in colour and has an earthy smell; it should be covered until used so that rain does not leach out the soluble nutrients
Source: FAO filmstrip Compost - Africa

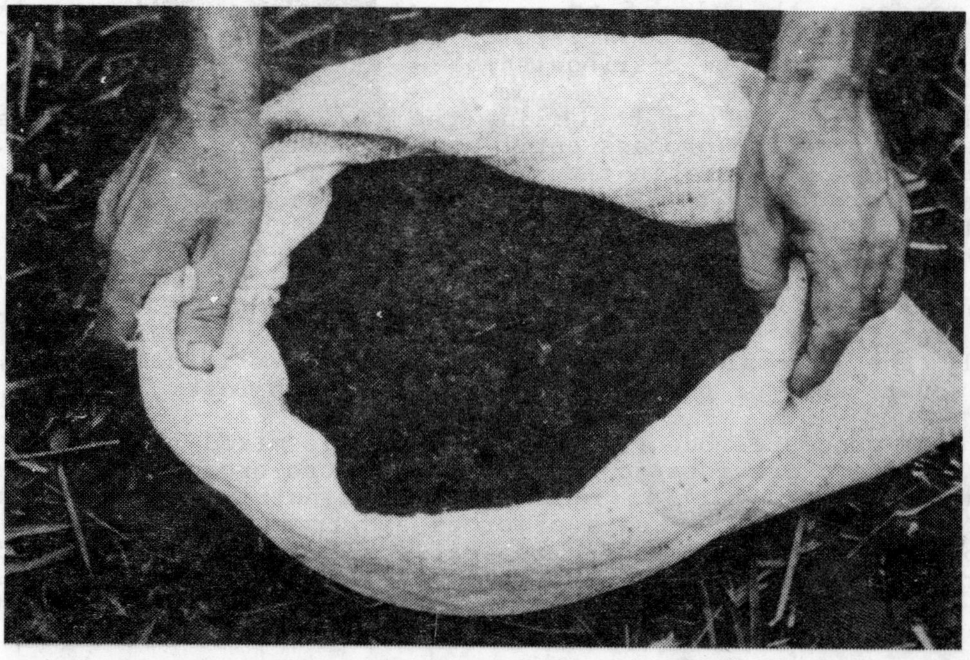

Plate 24 Sieved compost ready for use
Source: FAO filmstrip Compost - Thailand

supplies food for the micro-organisms already present in the soil. In their continued attack on organic matter the organisms help release gummy materials. In addition, the mycelia or hyphae of fungi and actinomycetes attach themselves to the soil particles as shown earlier in Figure 3. In this way large sand particles are held together into crumbs while fine clay and silt particles are opened out to provide channels for air and water movement and the growth of plant roots.

Some fungi in this situation form mycorrhizal associations with the roots of certain plants and trees; these 'bridges' are beneficial in the transfer of nutrients from the soil to the plant.

A very active population of soil micro-organisms of many different species is believed to help keep in check pathogenic disease organisms which would attack the plant. In addition, certain predatory fungi have the ability to trap and kill eelworms (minute nematodes) which often attack the root system of cucumbers, potatoes and tomatoes.

Some micro-organisms appear to be able to attack and break down, or at least modify, some of the agrochemicals such as herbicides and pesticides which are used in agriculture operations.

5.3.3 Chemical Aspects

Compost is prepared from essentially organic wastes. During the process much carbon, hydrogen and oxygen, and a little nitrogen, are given off into the air; otherwise the remaining elements are in the product. The chemical composition reflects the nature of the wastes from which the compost is prepared. Table 14 illustrates the wide range in chemical analyses which occur. Composts made from municipal wastes, refuse and night soil or sewage sludge, have compositions towards the left hand column of figures; these show low levels of organic matter and the major plant nutrients but high ash (mineral matter) and calcium contents. Farm and garden wastes produce composts with much higher organic matter and major nutrient levels. These analyses are on a dry weight basis; the moisture content of product composts in the tropics is probably in the range 35 to 55 percent on a fresh weight basis.

Table 14 CHEMICAL COMPOSITION RANGES OF MATURED COMPOSTS

Substance	Composition range weight % on dry basis		
Organic matter	25	to	80
Carbon	8	to	50
Nitrogen (as N)	0.4	to	3.5
Phosphorus (as P)	0.1	to	1.6
Potassium (as K)	0.4	to	1.6
Calcium (as Ca)	5.0	to	1.1
Ash (mineral matter)	75	to	20
	Municipal compost		Farm/garden composts

Table 15 shows the major nutrient contents of a number of composts but expressed on a fresh weight basis. Composts 1 to 5 were made under tropical conditions; the appreciable levels of potassium (K) are due to the very careful conservation of animal urine.

Unlike the case with water-soluble fertilizers, the major plant nutrients in compost are not immediately available for uptake by plants.

Nitrogen is closely bound up in the protein of the bodies of the micro-organisms and in the product humus and is only released on mineralization. Phosphorus and potassium are less tightly bound and hence are more available Under temperate climates the availability in the year of application is: nitrogen - 25%; phosphorus - 100%; potassium - 80%. Under tropical climates, with faster mineralization, the nitrogen availability is likely to be much higher.

Table 15 COMPOSITION OF VARIOUS COMPOSTS (EXPRESSED AS PERCENT OF FRESH WEIGHT)

Composts made from:	Composition %N	%P	%K
1. Cotton stalks and dung	0.40	0.13	1.4
2. Vegetable wastes	0.49	0.12	0.9
3. Vegetable wastes and dung	0.43	0.10	1.0
4. Mixed weeds and sunn hemp	0.41	0.11	1.7
5. Mixed weeds	0.40	0.12	1.3
6. Poultry dung and straw	1.1	0.9	0.8
7. Garden compost	0.4-3.5	0.3-1.0	0.2-0.3
8. Straw and sewage sludge	0.5	0.15	0.12
9. Municipal compost	0.4-1.6	0.1-0.4	0.2-0.6
10. Poultry manure and sawdust	1.0	0.4	0.46
11. Pig slurry and straw	0.53	0.37	0.33

Table 16 TOTAL TRACE METAL CONTENTS OF SOME U.K. MUNICIPAL COMPOSTS (EXPRESSED IN mg/kg ON A DRIED WEIGHT BASIS)

Trace metal	Content Range	Mean value
Arsenic (As)	23 - 74	52
Boron (B)	108 - 240	174
Beryllium (Be)	9 - 39	18
Cadmium (Cd)	2.5 14.8	7.5
Cobalt (Co)	6 - 84	48
Chromium (Cr)	100 - 280	170
Copper (Cu)	280 1100	610
Mercury (Hg)	15 - 25	21
Manganese (Mn)	385 - 1600	800
Molybdenum (Mo)	20 - 35	25
Nickel (Ni)	90 - 180	140
Lead (Pb)	385 - 4100	1630
Selenium (Se)	3.5 - 6.9	4.8
Titanium (Ti)	2200	2200
Vanadium (V)	38 - 310	170
Zinc (Zn)	465 - 2250	1350

Source: Gray and Biddlestone (1980)

In addition to the major and minor plant nutrients, composts prepared from municipal wastes, especially in heavily industrialized areas, often contain significant contents of trace metals. Data by Gray and Biddlestone (1980) on the trace metal content of several UK municipal composts are shown in Table 16; two of the composts were based on refuse only and two on refuse and sewage sludge mixtures. In vegetable growing trials on soils treated with very heavy dressings of these municipal composts, about 250 t/ha, it was found that the content of the trace metals in the soil was significantly raised. However, the trace metal uptake by the plants was not increased by anything like the same amount, and was sometimes actually reduced. This can be attributed to the ability of the organic matter to bind or chelate the trace metal into organic complexes and to the appreciable calcium content and alkaline pH value (averaging 7.4) of the composts; all these factors reduce the availability of trace metals to plants. In a similar manner, heavy dressings of organic matter can help to bind high concentrations of metals in soils contaminated by mining or industrial operations, rendering these areas less toxic to plants and thereby enabling vegetation to develop.

The addition of compost greatly increases the cation (or base) exchange capacity of the soil and can significantly improve the uptake by plants of nutrients from mineral fertilizers.

As compost breaks down, liberating carbon dioxide and moisture, it increases the concentration of carbon dioxide in the soil and just above the surface. This aids the photosynthesis of plant material and is believed to account for much of the increased yield when crops are heavily manured.

5.3.4 Physical Aspects

The addition of compost has a very beneficial effect upon soil structure and the properties related to it. The humus into which compost gradually changes is a colloidal material which, together with the gums and hyphae of certain micro-organisms, helps soil particles to bind into crumbs or granules with passages between them. Another contributing factor is that organic matter increases the number of earthworms which burrow through the soil. As these passages form, so the soil bulk density decreases while its porosity and hence permeability to air increase. The changes allow easier penetration of plant roots and percolation of rain water down through the soil. The stickiness of the soil is also reduced making it easier to work, thereby requiring less energy from human workers and less draught power from animals.

Organic matter can absorb considerable quantities of water, often 5 to 6 times its own weight. Hence the addition of compost will greatly increase the water retention in soils, holding up more in the crop root zone and allowing less to pass through into subsoil layers. Likewise it greatly reduces the rate of water evaporation into the atmosphere. These factors are of very great importance in areas where rain only falls on a few occasions each year and crop yields are highly dependent on the amount of accessible water in the crop root zone. Under conditions of very heavy rainfall the elasticity of a soil of high organic matter content helps to absorb the energy of falling rain drops so that surface damage is lessened. Soil erosion by water on steep hillsides is very much reduced; one of the current major markets for municipal compost in France and Germany is for use on steep vineyard slopes.

Although compost gradually supplies plant nutrients as it mineralizes, its major importance in tropical and subtropical environments is in its ability to improve water supply to growing crops and to stabilize soils against erosion by water and wind.

5.4 USE OF COMPOST FOR GARDENS AND MARKET GARDENS

For many people the garden is an area surrounding or close to the house where vegetables are grown intensively for use by the household. With slightly more land area, 1-2 hectares, some cereal crops can be grown and a few tree crops such as fruit or nuts. Occasionally there will be excess items for sale or trade. Flower growing and grass lawns for relaxation are usually a consideration secondary to food production.

Because the garden is close to the house the standard of care of the crops tends to be better than in the field. In addition water is more readily available. However, an almost continual supply of vegetables is expected, although this is not always possible to achieve. In many communities it is the job of the women to tend to vegetable gardens, fitting this in with their household tasks, although the men will often help with the work of soil preparation.

The main objectives and means to achieve them in garden cultivation are:

i. to grow as intensively as possible;

ii. to grow a rotation of crops which are likely to be successful and which will provide good nutrition for the household;

iii. to provide plenty of compost for its benefits to soil structure, supply of plant nutrients and moisture retention properties;

iv. to mulch the soil surface to reduce loss of moisture and oxidation of compost.

5.4.1 Setting up a New Garden

Many householders will have little choice over the site for their gardens and will have to use the land attached to their homes. This may range from a small backyard to a portion of a large area of communally-owned land. The size of the plot will depend on the availability of the land and the amount of time that the family has to work in it. With a reasonable climate, soil and access to water, an area of about 200 m^2 should provide the vegetables required by a small family for most of the year. The ideal site is a level piece of ground with good fertile soil close to a supply of water.

An adequate fence around the site is normally essential to keep out livestock. Fencing is expensive so if several gardens can be surrounded by one fence this reduces cost.

The plot then needs clearing to remove grasses, weeds and small bushes from the site. These wastes should be saved ready for making the compost heap. Considerable thought should be given before trees are removed as large-scale deforestation has caused so much trouble with soil erosion. Trees are useful for supplying wood, leafy branches for mulching, leaves for composting and shade for seed beds and compost heaps. They also bring up nutrients from subsoil layers which are recycled via their leaves and branches.

The compost heap should be sited and compost production started as described in Section 4.2, using wastes from the household, the garden and any locally available materials.

A decision has to be made about the bed system to be used for growing. The former practice of large rectangular beds with rows of vegetables has given way in recent years to more easily managed and more.

productive arrangements. With the original scheme, spaces had to be left between vegetable rows so that the gardener had access for sowing, planting, weeding and harvesting. These bare spaces involved wasted space, fertilizer and water.

5.4.2 Strip Beds

A much better approach is to use long narrow beds with paths in between. The beds are filled with vegetables growing closely together and the gardener can reach the whole area from the paths without treading on cultivated soil. A suitable layout is to have beds 1.2 to 1.6 m wide with paths 500 mm wide in between which allow access for gardeners and wheelbarrows.

If the garden is on a slope, the beds should be laid out along the contour, across the slope, to prevent soil erosion in heavy rain.

Where the land has been badly neglected it is usually advisable to dig deeply, loosening the soil to a depth of at least 500 mm, particularly on heavy clay and silty soils. Once the garden has been established, deep digging should not need to be repeated because normal gardening operations with regular additions of compost will keep the soil in good condition, particularly if earthworms can be attracted into the area.

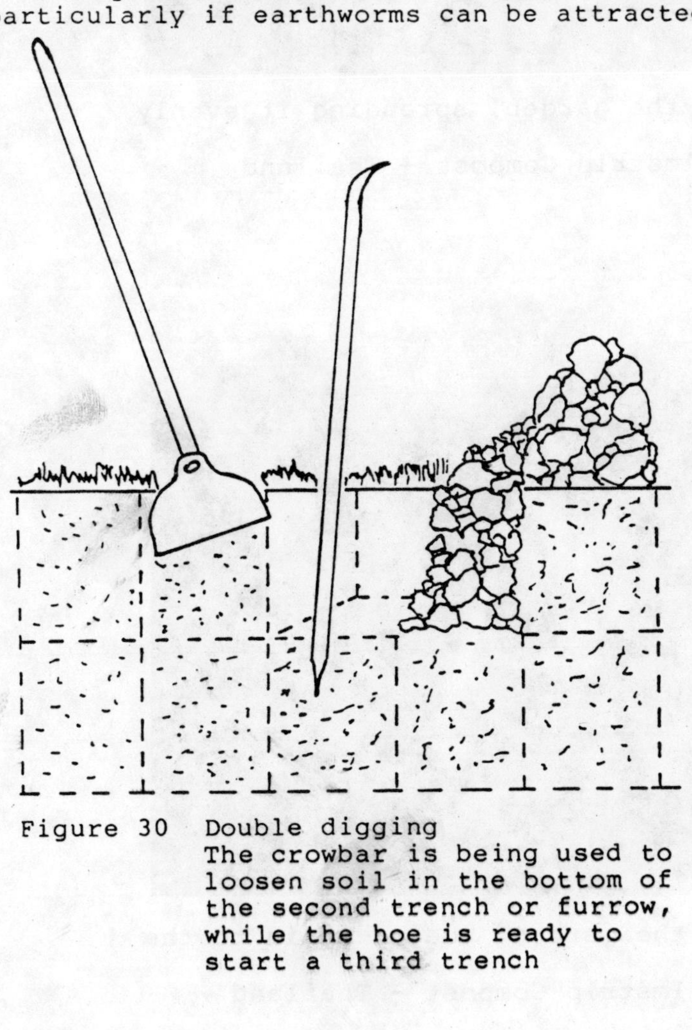

Figure 30 Double digging
The crowbar is being used to loosen soil in the bottom of the second trench or furrow, while the hoe is ready to start a third trench

One method of loosening the soil to a depth is 'double digging'. A trench 250 mm deep is dug along the end of the strip bed and the soil removed. The soil underneath is then loosened with a crowbar or other implement, but this bottom layer is not lifted out or turned over. A second trench alongside is then dug out, the soil being used to fill the first one. Again the lower layer is opened up with a crowbar. Then a third trench is dug and the process repeated along the strip bed. Finally the soil removed from the first trench is used to fill in the last one to be dug. This method is shown in Figure 30. It is important not to bring up material from the lower layer into the top, as subsoils in the tropics are often very infertile.

When carrying out such double digging, compost should be mixed in with the top layer of soil if the bed is to be used for growing crops in the near future (Plates 25 & 26).

Plate 25 Compost use in the garden, spreading it evenly over the soil
Source: FAO filmstrip Compost - Thailand

Plate 26 Compost use in the garden, hoeing it in to the soil
Source: FAO filmstrip Compost - Thailand

If sufficient compost is available, heavy dressings of at least 100 tonnes/ha (1000 kg/100 m^2) will greatly improve fertility. If the bed is not to be sown or planted immediately it is best to leave the addition of compost until shortly before sowing or planting takes place. As stated previously, compost for use with sowing small seeds must be fully matured; that used for mulching around established plants, shrubs and trees need not be so well decomposed.

5.4.3 Fertility Trenches

This method is a small-scale version of the Mazibuko system described in Section 4.3.3. It is ideal for small gardens where compost heaps would take up valuable space. In heavy downpours of rain the trenches collect surface run-off water and allow it to soak down readily to the rooting zone of the crops.

The trench is dug about 1 m wide and 500 mm deep. The topsoil from the first 150 mm depth is set aside for covering up the trench later. The remaining soil is used for building up the paths between trenches so that sunken beds are formed. The soil in the bottom of the trench is loosened with a crowbar or other tool.

The trench is then filled with as great a variety of organic wastes as are available from the home, garden and neighbourhood. Thick layers of newspapers or grass cuttings should be avoided as these tend to form wads and do not break down properly. A little soil is added and the whole mass mixed together and well watered. The organic material should be pressed down with a hoe but not compacted too tightly. The topsoil, which was set aside originally, is then replaced. The result will be a slight mound; the organic material underneath will gradually break down and sink. Ideally, the bed should be left for a few weeks before use to allow the composting reaction to get under way and major settlement to take place. If insufficient organic waste is available to fill a complete trench, the latter is constructed, filled and completed in short sections.

5.4.4 Mound Cultivation

The intensive cultivation of vegetables on raised beds or mounds is a practice being used in many parts of the world; significantly improved yields of crops are claimed. It is not so suitable where torrential downpours of rain occur unless the soil is well protected against erosion.

As shown in Figure 31 the mound is built up of several layers; the bed length is dictated by the amount of organic material available. Several mounds are often built, with pathways in between. The life of a mound is about 5 years by which time it has sunk to ground level; it is then rebuilt.

In building such a bed, a trench is dug about 1.5 m wide and 300 mm deep; the soil is set aside for use later. Then woody plant material, twigs, branches, maize and sunflower stems, are laid out in a rounded heap about 550 mm high down the middle of the trench (Figure 31); at each end a space of nearly 1 m is left so that the mound can be sloped down. This central core is then covered with a layer of grass turves, placed face down, with a little soil to fill in any gaps. The next layer, 300 mm thick, is of moist leaves, if possible from a variety of trees. This is covered with coarse immature compost about 150 mm thick. The mound is then completed with 150 mm of finely sieved, fully matured compost mixed with topsoil. The various layers are well watered as construction proceeds.

The various vegetable crops are normally sown on the contours of the mound as shown in Figure 31, with tall plants such as tomatoes planted along the top.

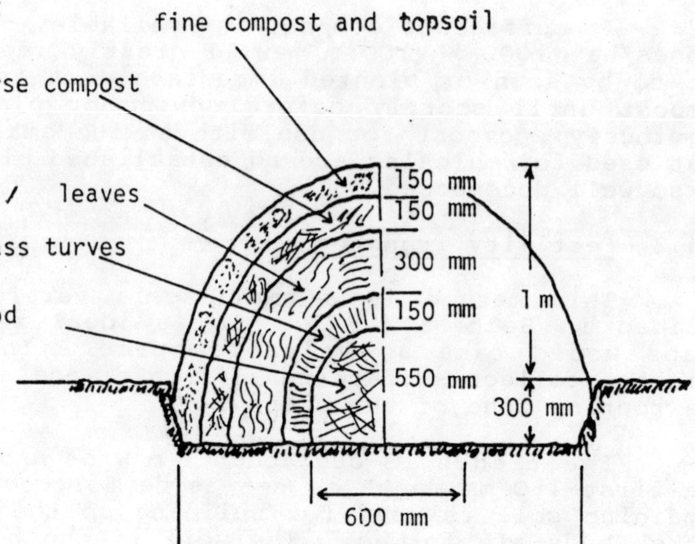

Instructions for building a mound

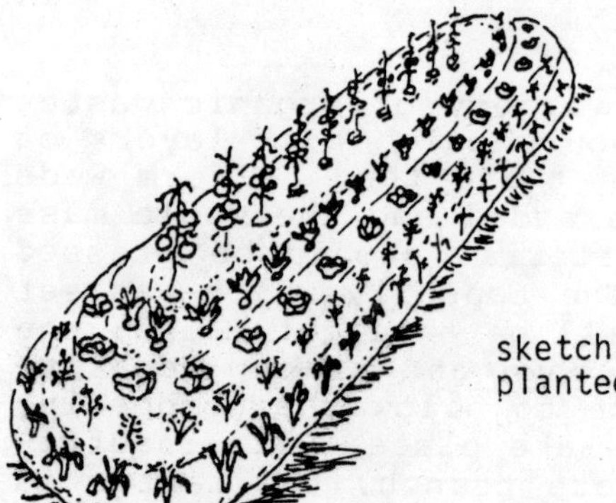

sketch of a planted mound

Figure 31 Mound cultivation

5.4.5 Mulching

Mulching the soil to reduce moisture evaporation and decomposition of organic matter in strong sunlight is most important in gardens in tropical climates. In addition, it reduces caking and cracking of the soil, prevents erosion in heavy rain, and smothers weed seedling growth thereby virtually eliminating the need for hoeing.

Mulches can be of fine grasses, leaves, leafy branches, straws, stalks or coarse compost. They are very attractive to earthworms and on decomposition will be drawn down into the soil; they will then need renewing. A thick mulch, 50 to 150 mm deep, will considerably decrease moisture loss from the soil surface and therefore reduce the need for watering. This is because it lowers the temperature of the soil surface and also covers the capillary passages through which water rises to the surface. However, the soil underneath the mulch should be examined periodically to see when watering is needed.

When organic material is in short supply then mulches of pebbles and flat stones can be used, particularly around trees. It also helps if the soil surrounding the tree is sloped to form a shallow basin or depression with the tree at the centre; in this way water from any light falls of rain is guided to where it is needed and not wasted.

Another useful technique is to coat young vegetables such as leeks with a clay-compost mixture on transplanting from nursery beds. This mixture is made from equal weights of fine compost, clay and water. The roots of the leeks are trimmed to 10 mm and the stems/leaves to 100 mm; they are then dipped in the paste before planting out in their final positions. This appears to reduce the plants' need for water at this critical stage.

5.4.6 Market Gardens

These are commercial operations generally sited on the outskirts of a town or city. The produce grown is mainly vegetables, a small but increasing quantity of flowers, and some fruit; it is usually sold for cash to the urban population. The site is normally well chosen with fertile soil, access to an assured supply of water and adequate transport routes for outgoing produce and incoming raw materials.

The land is worked very heavily, several successive crops normally being grown each year on every piece of land; probably not so much care can be taken in protecting the soil as in a household garden. In contrast, some purchases of mineral fertilizers, organic manures and composts can be afforded. Much heavier and more frequent applications of compost are required than for field crops; inputs possibly total at least 100 t/ha/year. The compost may need to be produced quickly, so turning may be required and possibly the addition of nitrogen and phosphorus mineral fertilizers. Some organic wastes for composting will be available from the market garden itself, such as vegetable trimmings, roots, stems, stalks and unsaleable produce.

Being close to centres of population market gardens may have access to market wastes which will be of very high vegetable and putrescible content. In many countries throughout the tropics market gardens surrounding cities send in their produce by cart and bring back market wastes. Such wastes break down rapidly, supplying plant nutrients, but may add little to the long-term humus content of the soil. In addition to market wastes, urban refuse from the streets may be brought back for composting.

In parts of West Africa slightly decomposed or even fresh refuse transported in municipal lorries is a very popular manure (Grubben 1982). Applied at rates of 100 to 200 t/ha/year such materials can raise the fertility of new plots of poor sandy soil within a few years. The main vegetable type is amaranth, a form of spinach; this grows well on such fresh refuse. Some growers dig in up to 200 kg of refuse per bed of 10 m^2 (200 t/ha) and then raise 3 or 4 successive crops of the vegetable before applying further fertilization. Other growers plant vegetables such as cabbages and French beans after the initial amaranth crop; by then the decomposition of organic matter in the soil has settled down. The results of experiments with growing amaranth showed that the local refuse with its high vegetable/putrescible content is a good organic fertilizer for this vegetable; no set-backs to growth due to nitrogen robbing were observed and no hygienic problems were encountered.

Such good results cannot always be expected, especially with refuse containing a higher paper content and which has not been partly composted while lying in heaps in the streets; nor with all types of vegetables. The present authors would prefer to see some degree of composting to ensure hygienic conditions. The other problem is the virtual impossibility of removing all non-degradable material such as crockery, glass, metal and plastics from raw refuse. In our own experiments on vegetable growing using heavy dressings of several U.K. municipal composts, one compost was poorly sieved; as a result the cropping area soon gathered an unsightly amount of rubbish on the surface.

5.5 USE OF COMPOST FOR TREE PLANTING

The reclamation of desert land and the improvement of agriculture on eroded and infertile soils will depend to a large extent on the ability to re-establish trees. Large-scale deforestation this century has led to severe soil degradation problems in several countries. Compost can make a significant contribution to the raising and planting out of small trees and in their subsequent care.

5.5.1 The Uses of Trees

Trees are important for many reasons:

i. protecting soils against water and wind erosion;

ii. providing boundaries for fields and fencing for livestock;

iii. supplying firewood on a carefully organized programme of cutting;

iv. supplying crops for food and sale, and also timber;

v. providing leafy branches for mulching soils and leaves for composting;

vi. providing shade;

vii. bringing up plant nutrients and moisture from deeper soil layers;

viii. modifying the local micro-climate by cooling the temperature, raising the humidity and eventually giving rise to dew.

Plate 27 Alley cropping. Young maize growing in alley formed by Leucaena leucocephala; leaves and twigs are used for mulching and the wood piled at the back is from the stems
Source: Wilson and Kang (1981)

5.5.2 Alley Cropping and Forest Farming

Alley cropping is an agricultural practice with much potential in which food crops are grown in alleys 4 to 6 m wide formed by fast growing shrubs and trees planted in an orderly arrangement (Wilson and Kang 1981). Leguminous shrubs and trees are employed such as *Glyricidia* species, *Leucaena leucocephala*, the pigeon peas *Cajanus* species, and *Tephrosia candida*. These supply nitrogen to the soil via their root systems while green leaves, often amounting to 12 - 16 kg per year per tree, and twigs serve as mulch and a supply of further plant nutrients (Plate 27). The leaves can also be used for composting. The larger branches are used for poles, stakes for climbing plants, or as a source of firewood. In the growing season the trees are kept cut and pruned to avoid competition with the arable crop; then in the dry season the trees are allowed to grow, drawing their moisture and nutrients from the deeper soil levels. The benefits of the system improve with time. In experiments in Africa on originally infertile sandy soil, alley cropping is giving maize yields far higher than the average.

Forest farming is a more integrated combination of agriculture and silviculture (forestry) in which blocks of trees are planted alternately with arable strips used for pasture or cereal growing. It is described in detail by Douglas and Hart (1976).

5.5.3 Tree Raising, Planting Out and Manuring

The use of compost plays a part in all these aspects of tree care.

Young trees are normally propagated from seeds or cuttings in a nursery. Until recently the practice has been to use nursery beds with soil of firm tilth on which well-sieved compost or leaf mould has been spread and raked in. Under tropical conditions both wind breaks and careful shading against the sun are very important to protect the plants and reduce moisture loss from the soil. At an appropriate stage of growth the young plants are lifted from the nursery bed in the dormant season and planted out into their final positions. Unfortunately, the loss of soil from the root system has often proved too much of a shock on planting out and plant mortality has been high.

A more recent practice is container growing. Very young seedlings or rooted cuttings are transplanted into containers which are filled with a mixture of half compost and half soil. On commercial plantations containers made from polythene plastic tubing are used; in other cases containers may be made from grasses, rushes or leaves sewn together. Even tins, with drainage holes punched in the bottom, can be used. The trees are grown in these containers until they have reached a suitable size for planting out. Use of such containers reduces tree mortality and means that planting out can be done throughout most of the year, except for the hottest months.

The holes for planting should be dug out and prepared carefully. Ideally this should be done well before planting to allow weathering and soil sterilizing in hot weather; fungal infections and termite activity will then be greatly reduced. For most trees the maximum hole size is approximately 800 mm square and 600 mm deep, though this will vary with the type of tree. For eucalyptus seedlings a hole 300 mm square is sufficient. Holes of too small a size cause stunted growth of plants, particularly in poor soils. Square holes are important where the ground is very hard or rocky, otherwise the roots tend to grow round instead of striking out into the surrounding soil. The topsoil removed is kept apart from the subsoil. The base of the hole is then broken up with a crowbar or spike.

Planting is normally done at the start of the rainy season unless irrigation is available. The topsoil is mixed with at least 6 kg of compost and the hole partly filled (Plate 28). The plant is then set in the hole, making sure that plastic or metal containers are removed first; ones of grass or leaves can be left around the plant as they will break down and allow the roots to pass through. Filling of the hole is then completed, firstly with the remains of the topsoil-compost mixture, then with the subsoil which is made firm by treading down. Use of compost plus all the excavated soil will result in a slight mound around the tree seedling; this will prevent waterlogging in the rainy season. Immediately before the start of the hot weather the soil at the base of the tree is hoed to control weeds and mulched with coarse compost to prevent the soil drying out.

In arid areas a shallow depression or basin is left at the top of the planting hole which is then well watered. The depression is then filled with a thick mulch of organic material if available or small stones (Figure 32). Such mulching is most important as it covers up the capillaries or narrow passages through the soil by which the soil water rises to the surface and evaporates.

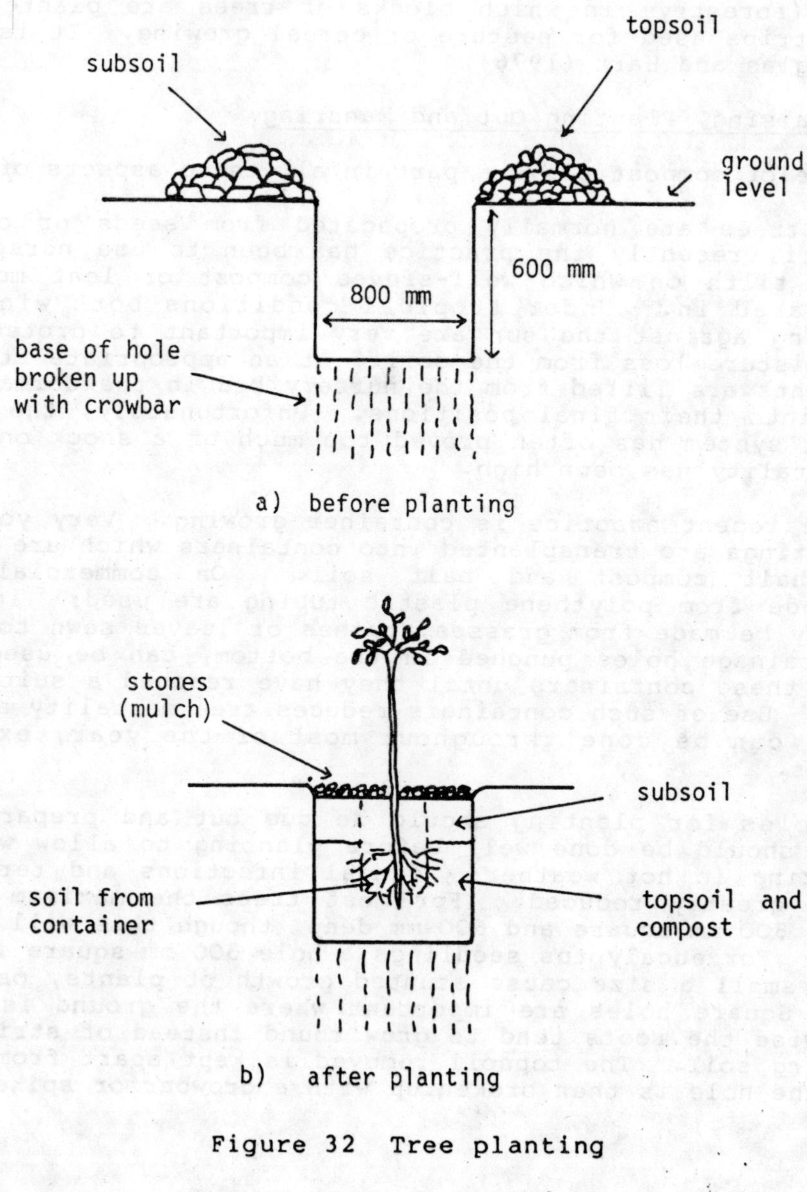

Figure 32 Tree planting

Plate 28 Tree planting: mixing compost with topsoil ready to plant the tree from the container
Source: FAO filmstrip Uses of Coffee Pulp

Plate 29 Use of compost on established trees; a shallow trench is dug well away from the trunk, the compost put in and then covered over
Source: FAO filmstrip Compost - Thailand

On steep hill slopes trees should be planted in trenches dug on the contour, similar to the Mazibuko system of Section 4.3.3. The trenches act as an erosion control and help the rainwater soak down to the root zones of the trees.

When a tree is established and growing well it benefits from periodic dressings of compost. As shown in Plate 29, it is spread over the feeding root zone. A shallow circular trench is dug with a mattock, the compost put in and then covered with soil.

For bushes such as coffee about 6 kg of compost is added into the planting hole, then annual dressings of 10-20 kg given as the bush comes into production. Guava and limes can be given 15-20 kg of compost annually. Oranges benefit from 20-25 kg in the planting hole with annual dressings per tree of the same amount before flowering, building up to a maximum of 50 kg when in full production. Pineapples can be given 25 kg at planting, then further dressings of 25 kg every 6 months. For bananas, 25 kg of compost per plant can be used at planting out.

In some plantations where controlled heap or pit composting is not practised, the organic waste materials are placed in trenches dug between the rows of shrubs or trees. The roots then grow out into these trenches and benefit from the nutrients released on breakdown of the material. This technique is practised on coconut, rubber and tea plantations.

5.5.4 The Mycorrhizal Association

In plantations in which a high concentration of organic matter is maintained, the soil contains many micro-fungi and their hyphae or mycelia. In such situations many shrubs, trees and plantation crops have been shown to form mycorrhizae. These are symbiotic associations between the plant root and the soil fungus which contribute to the health and nutrition of the crop. The fungal hyphae contain significant levels of nitrogen in the form of protein which is digested by the plant roots. Mycorrhizal association has been found in banana, cocoa, coconut, coffee, cotton, rubber, sugarcane and tea plantations.

5.5.5 Desert Reclamation

The Sahara Desert in North Africa is still extending southwards at the rate of several km each year. Severe drought conditions in 1984/85 have brought famine to many millions and threaten the livelihood of an estimated 150 million people. Although some progress on desert reclamation has been made by some of the countries on the northern edge of the Sahara, little has been attempted on the southern edge; the scale of the problem is now vast. In addition to deserts there are many badly eroded hilly slopes which are urgently needed for food production by expanding populations.

Desert reclamation can be achieved by the stabilization of sand dunes against wind and water by planting tree belts. Compost has a role to play by its ability to hold moisture in the immediate root zone of the newly planted tree until the latter can extend its own roots to seek out moisture held at depth in the soil.

Initially, very drought-resistant species of trees are required. They are planted from containers in a manner similar to that shown in Figure 32. However, better results are obtained by producing long-stemmed seedlings in which the roots are planted about 500 mm below the soil surface; the root zone is surrounded with the compost-topsoil mixture, the hole filled up with subsoil and watered. The depression at the top is covered with a thick mulch of stones and, if sufficient labour is available, the local soil surface is sloped slightly to form a micro-catchment area to direct rainwater to the tree. Such plantings of young trees need protection from wind-blown sand and browsing animals, especially goats. When such trees survive planting out they quickly put out a very extensive root system, some species reaching down at least 50 m, to seek out moisture from the subsoil.

Once the drought-resistant trees have become established, the soil stabilized and the micro-climate improved by an increase in humidity in the local area, then more valuable trees supplying food, fodder and fuel can be planted. Large-scale monoculture of trees should be avoided. Advice on the best tree varieties and combinations to grow for a particular purpose should be sought from the local agricultural or forestry extension officer.

In a 5-10 year programme the amount of vegetation can be enormously improved, food production restored and the area repopulated. The major problems at the start are the supply of fencing material, the availability of compost and the close care of the young trees. The initial establishment of trees in an area may well require the use of expensive fencing; however, once these first trees are growing they will provide shelter against the wind and some fencing material for subsequent plantings. Shrubs to provide material for local compost production can be planted among the initial trees. Many tree planting programmes have achieved poor results, often due to destruction by goats, particularly where the trees were regarded as belonging to 'the government' and looked upon as a source of free fodder. In one area success only resulted when blocks of trees were given to families who, in return for care of the trees, were allowed to keep the resulting crops, fodder and firewood. This question of social responsibility will undoubtedly involve the issue of land ownership. Without solving this problem, desert reclamation may well prove impossible.

5.6 USE OF COMPOST FOR FIELD CROPS

Compost should be applied as close to the time of crop establishment as possible. This will avoid losses of nutrients due to exposure to sunlight and will enable young plants to take advantage of the flush of nitrogen and phosphorus mineralized in the soil as a result of the input of compost. To obtain a worthwhile response it is advisable to apply compost at a minimum rate of 7.5 t/ha. It is difficult to spread smaller quantities evenly on the land. The application rate chosen will obviously depend to a great extent on the amount of compost available and in practice it is worthwhile to aim for dosage levels of 25-40 t/ha. Assuming there is sufficient compost available to achieve the minimum rate, it should be applied evenly all over the land. If smaller quantities are available, it is better to spread it on part of the land only at the minimum dosage rate and then to rotate dressings around the land in successive plantings. Some examples are given in Table 17.

Table 17 EXAMPLES OF COMPOST APPLICATION

Compost available tonnes	Total land area hectares	Suggested usage t/ha
15	2	7.5
30	2	15
15	4	7.5 on 2 ha only; remaining land to be manured for next crop

If compost is prepared at one time of the year, it is probably best to use it all for the next crop unless dressings of more than 60 t/ha are available. In this case, it may be better to split the dressing between two succeeding crops.

When compost is applied to land it can be left on the surface as a mulch or ploughed in before the crop is planted. With low rates of application the mulching effect will not be good and hence it is better to plough the compost in to a depth of not more than 100 mm. With very high rates of application ploughing in may not have any advantage over mulching. Mulches have been shown to have similar properties to soil organic matter in reducing erosion, lowering soil temperature, increasing the infiltration of rain and lowering evaporation losses from the soil. An additional advantage claimed for mulches is that they reduce weed growth.

In some areas plastic sheet and petroleum by-products are used as mulches. They solve the problem of huge quantities of organic wastes required for large-scale mulching. While reducing evaporation and erosion, they can reduce the amount of water entering the soil and, therefore, are not as good as crop waste mulches. Due to their high cost it is unlikely that they will be economic except for high yield and high value crops. A viable alternative practised in several semi-arid areas is the use of small pebbles to cover the soil between crop rows and around trees. Pebble mulches protect the soil from high impact rainfall and help to lower soil temperature and water evaporation. They also increase the infiltration of water.

If immature compost is applied as a mulch there is a danger that the continuing breakdown process may release products into the soil which inhibit small seed germination. Provided that the compost is mature it is unlikely that a harmful quantity can be applied. This is a benefit of using compost rather than manures or undecomposed wastes which can often be harmful to crops.

5.7 IMPROVED CROPPING TECHNIQUES

The purpose of producing and using compost is to improve soil fertility. If this is to result in the maximum possible stable increase in crop production, it is important that other aspects of the cropping system should also be optimum. The following improvements to cropping techniques have given good field level results in many areas and are worth consideration in any programme for raising yields (Dalzell 1981; FAO 1984a, 1984b and 1985a).

5.7.1 Intercropping

Traditionally, intercropping has been widely practised in tropical agriculture. Several types of crop have been grown simultaneously in one field. The main benefit seems to have been an insurance against failure of any one crop. If early rains were poor the quickest growing crops usually failed but the slower crops had a chance of responding to later rains and vice versa. This insurance could be obtained by growing several crops separately. It is now known that intercropping can improve production levels over sole cropping. Thus two hectares intercropped with maize and pigeon pea can be expected to yield more of each crop than one hectare cropped with maize only and one hectare cropped with pigeon pea only (Figure 33). Among the reasons for the yield advantages obtained from intercropping are the following:

i. the different branching and leaf patterns of the various crops intercept more light than do single crops;

ii. the roots of the different crops use different zones of the soil; hence more nutrients and moisture are utilized;

iii. different crops grow at different rates and thus a slow growing crop will not compete in the early stages with a quick one. The quick growing type uses moisture which would be lost if only a single slow growing crop was in the field. The slow grower in its later stages, after the harvesting of the quick grower, uses soil moisture which would otherwise be lost;

iv. pest, disease and weed problems are reduced;

v. legume crops in an intercrop produce nitrogen for other crops. Originally this was thought to be the main benefit but it is now believed to be one of the least important.

For maximum efficiency it is important to use the optimum plant populations and spacing. Crops which interact well together should be sown and the choice of variety is especially important. Further details can be found in the publications of the International Agricultural Research Institutes and from local research stations and government advisory services.

5.7.2 Single or Fixed Row Cropping

In infertile rainfed soils benefits can be obtained from cultivating and cropping on the same rows in successive seasons. Crops are sown in straight lines and before the succeeding crop is planted only the rows of stubble are cultivated. The new crop is sown in these same lines and all fertilizer, mineral or organic, is put into them. As a result the crop lines become areas of concentrated fertility. The inter-row areas are cultivated only by shallow hoeing for weed control and the necessary soil structure because improved fertility slowly builds up in these areas. Repeated intensive cultivation over the whole field would prevent this natural build-up of fertility.

5.7.3 Paired Row Cropping

In this system, instead of having equal spaces between crop rows, two rows are grown close together with a wider gap separating them from the next paired row (Figure 34). The wide space between rows allows efficient hoeing for weed control; fertilizers can be applied to the paired rows, resulting in little wastage. The system is suitable for crops such as wheat and finger millet which are usually grown with narrow row spacings.

5.7.4 Broad Based Beds and Furrows

In many areas of the tropics drainage is a major problem of rainfed crops on black soils. The deeper the soil the greater the problem. As a result, many of these quite fertile soils are left fallow during the rainy seasons; crops are then grown in the dry season using moisture remaining in the soil after the rain has stopped. The formation of broad based beds 200-250 mm high at 1.5 m intervals with channels between them (Figure 35) has been found to prevent waterlogging on such soils and to make cropping possible during the rains. A sole crop of short duration can be followed by the traditional dry season crop. An alternative is a cereal-pigeon pea intercrop with the pigeon pea flowering towards the end of the rains and maturing with the residual moisture. In each case cropping intensity is greatly increased.

The raised beds, which are permanent in nature, are made along the contour so that they help to prevent run-off during light rains. Excess moisture in heavy rains can collect in the channels and be led off slowly into controlled waterways for drainage. The system is suitable for a wide

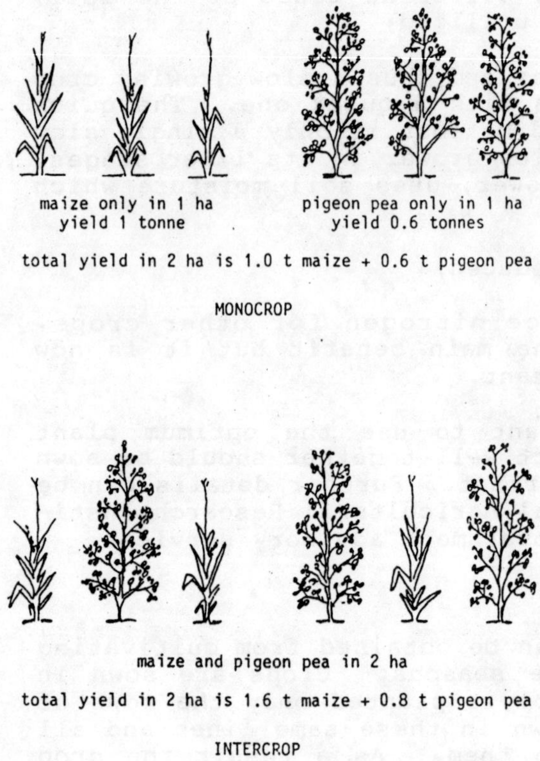

Figure 33 Advantages of intercropping

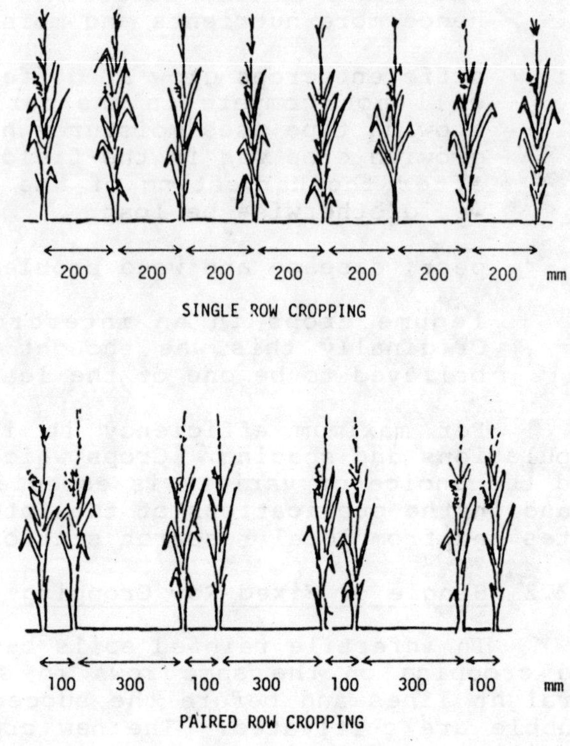

Figure 34 Single and paired row cropping

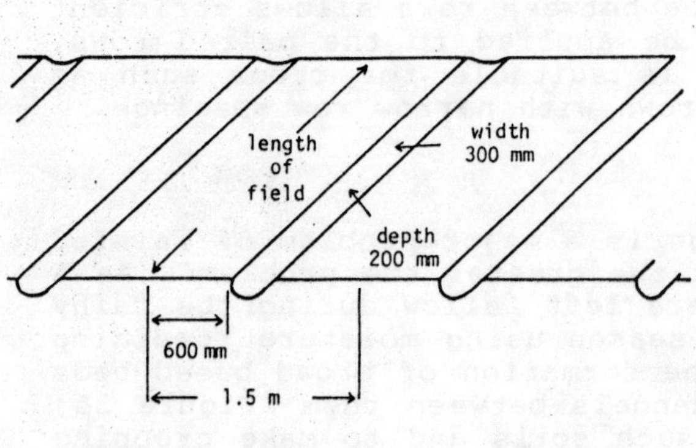

Figure 35 Broad based beds and furrows

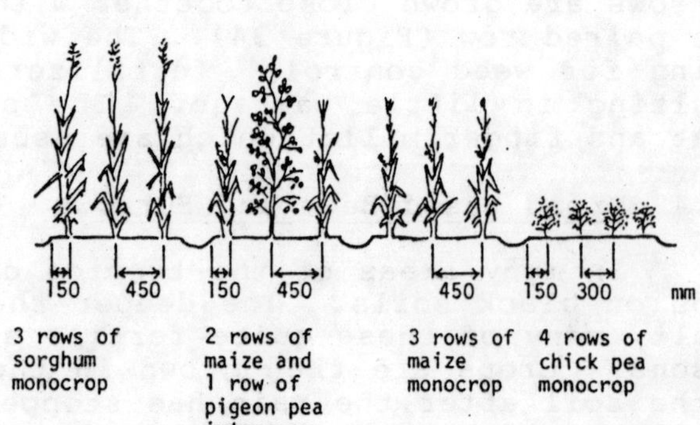

Figure 36 Crops on broad based beds

range of row widths and for intercrops with the cropping area usually being restricted to a zone 1 m wide centred on the middle of the bed (Figure 36). All fertilizers and manures can be applied in this zone and only minimal weed control cultivation need be done in the areas between cropping zones. Good soil structure develops in the beds and hence only limited cultivation is required to prepare seed beds. Yields of up to 5 t/ha/year have been achieved with this system under semi-arid rainfed conditions when growing maize in combination with chick pea, red gram or sorghum. Suggestions for further reading on this subject are given in Appendix 7.

5.8 ROTATIONS

Whichever pattern of cropping is employed it is useful to change the crop from season to season to prevent a decline in soil fertility. Different crops have different nutrient needs and different rooting patterns. Legumes return to the soil nitrogen which is required by cereals and root crops. The change of growth and cultivation patterns with different crops helps to prevent the build-up of weeds, pests and diseases which tends to occur with repeated monocropping. Apart from beneficial effects on soil fertility and pest problems, rotations also tend to even out labour peaks and economic risks. An example is given in Table 18 of rotations which could be used for farming in semi-arid rainfed conditions. The rotation involves intercrops and could be carried out in flat fields or, in the case of black soils, on the broad based bed and furrow system.

Table 18 ROTATIONS SUITABLE FOR SEMI-ARID RAINFED SOILS

	Red soils	Black soils
1st Year		
Rainy season	Maize + pigeon pea intercrop	Maize
Dry season		Chick pea
2nd Year		
Rainy season	Castor or groundnuts	Mung bean
Dry season		Sorghum
3rd Year		
Rainy season	Sorghum + pigeon pea intercrop	Maize + pigeon pea intercrop
Dry season		
4th Year	As for 1st year	As for 1st year

5.9 GREEN MANURING

Green manuring is the production of a short duration crop, usually a legume such as sunn hemp, which is ploughed into the soil before it reaches maturity. As it decomposes in the soil it provides nutrients for the ensuing crop. Initial cultivation done for the green manure helps to control weeds in the following crop. In rice growing areas, green manuring is a useful practice as the green crop can be grown between rice being sown in the nursery and transplanting it into the field. The succulent green crop will decompose rapidly when incorporated into the soil at the time of puddling. In this system there is no loss of time or moisture, whereas in rainfed cropping green manures would delay planting dates and utilize moisture that could otherwise be used by a productive crop. With increasing pressure on crop lands it would seem that the practice of green manuring will become less common. It is possible to adapt the production pattern of green manures to avoid competition with growing crops. Thus perennial shrubs such as Gliricidia can be grown along roadsides, around manure pits and on field borders, and the leaves spread on the fields and ploughed-in prior to sowing. Outside the cropping seasons the leaves can be used for composting.

5.10 DIRECT INCORPORATION OF MANURES AND WASTES

When considering this subject it is important to make a distinction between rice production under flooded conditions, crops which are given periodic irrigation and rainfed crops. There is evidence to show that paddy straw directly incorporated into paddy fields in the puddling operation gives better results than does compost prepared from paddy straw. It has been suggested that the availability to the rice crop of silica from the decomposing straw increases the yield. It is important to remember that it is relatively easy to incorporate straw into a flooded field with puddling equipment and that breakdown will be reasonably fast. Normally one week will elapse between incorporation and transplanting and the plant introduced to the soil will be a vigorously growing one at least 21 days old and with a developing root system. There is no question of a soil moisture shortage.

This flooded situation is quite different from the rainfed one where incorporation of straw into dry and probably compact soil is very difficult; straw breakdown will be slower than in flooded conditions and will start with the onset of rains. The maximum demand for nutrients for breakdown will be in direct competition with seed germination and crop establishment. Under these circumstances, together with the sloping nature and moisture shortages characteristic of most dry lands, preparation and incorporation of compost is a much better practice than straw incorporation.

Periodically irrigated crops fall into a middle position. The land is more likely to be reasonably level and therefore less susceptible to water erosion. The soil is unlikely to be so compact and hence incorporation will be less difficult than on dry lands. Direct incorporation of dung will be possible more easily than on dry lands but straw incorporation and decomposition will be problematical. When direct incorporation of dung or straw is possible without undue difficulty or possible harm to the crop, it has the advantage of being less costly than composting. However, it must be remembered that dung is a very valuable raw material for composting in that it can be used to balance the high C/N ratio of very coarse wastes which can thereby be composted and brought into a useful form. If the dung is incorporated directly then the coarse material will be more difficult to compost and may not be recycled at all.

5.11 COMBINED USE OF MINERAL FERTILIZERS, ORGANIC WASTES AND COMPOST

It must be accepted that each of the above products has drawbacks as well as advantages. These are listed in Table 19. The combined use of mineral and organic fertilizers has been shown to overcome most of the individual drawbacks and in some cases to enhance the advantages.

Table 19 COMPARISONS OF MINERAL FERTILIZERS, ORGANIC WASTES AND COMPOSTS

Material	Advantage	Disadvantage
Mineral fertilizers	1. Convenient	1. Easily leached
	2. Transport and handling costs may be lower	2. Continuous use may lead to breakdown of soil structure
	3. Quick crop response	3. Supply major nutrients only
	4. Concentrated nutrient source	
Organic wastes	1. May be cheaper than mineral fertilizers or composts	1. May be difficult to incorporate into soils
	2. Improve soil fertility	2. If high C/N ratio they will decompose slowly and may rob soil N
		3. If low C/N ratio they have little effect on soil structure
		4. Dilute nutrient source
Composts	1. Improve soil structure	1. Dilute nutrient source
	2. Control erosion	2. High labour usage
	3. Supply wide range of nutrients	3. Transport costs more than mineral fertilizers
	4. Hygienic disposal of pathogenic wastes	
	5. Reduce transport needs of bulky wastes	

5.11.1 Organo-mineral Fertilizers

Many crop wastes, such as wheat chaff, paddy and groundnut husks, and tailings from millets, are easily incorporated into the soil but then decompose very slowly and compete with the crop for nitrogen. Such wastes can be treated with ammonium sulphate to bring the C/N ratio below 10/1 and raise the nitrogen content above 2.5 percent on a dry weight basis. The mixture will then decompose rapidly in the soil without harm to the crop and without the labour cost of composting. The basis of the calculation of the nitrogen requirement is shown in Appendix 2.

Much work has been done in India (for example Dhar, Bose and Gaur, 1955) on the breakdown of paddy straw and other organic wastes in the presence of added phosphorus. The latter has been used in the form of rock phosphate, basic slag and superphosphate. The research applied both to the composting of the wastes and to their breakdown in the soil on ploughing-in. Good results were obtained by the addition of about 4.4 kg of P per tonne of straw. This results in faster breakdown and a much increased nitrogen content of the compost. When straw plus phosphate was ploughed-in there appeared to be no robbing of nitrogen from the soil; this mainly was due to stimulation of the nitrogen-fixing bacteria Azotobacter. This technique has been proven in experiments elsewhere. Unfortunately, basic slag is often not available now due to changes in steelmaking practices.

Claims have been made that coating superphosphate fertilizer with cowdung brings about a major increase in the response of maize to superphosphate. There is much practical experience in India to show that the response of rice to top dressings of urea is enhanced if the urea is mixed with small balls of mud which are pressed into the soil. The use of straw plus urea to give a combined C/N ratio of 12/1 has been shown to be comparable to lac-coated urea and sulphur-coated urea as a slow-release source of nitrogen. This straw-urea mixture left more nitrogen in the soil under flood conditions than split applications of urea.

5.11.2 Combined Application of Compost and Mineral Fertilizers

Many experiments carried out in India and China have shown better crop yields when a combination of compost and mineral fertilizers was used. Poorer results were obtained if double dressings of compost alone or mineral fertilizers alone were used. Advantages have been found with a wide range of crops such as cotton, maize, rice, sorghum, sugarcane and wheat. Similar results have been obtained in Bolivia where potato yields, which had been decreasing under mineral fertilizer applications, have started to increase with combined applications of mineral fertilizer and composts. Similar combined applications have resulted in better juice quality in sugarcane and in higher dry matter yields per plant in tomatoes.

It is possible to argue that these results can be explained mainly by differences in the quantities of nutrients supplied, as the organic fraction will always contain trace elements not found in the mineral fertilizers. While this may be largely true, it is not an argument for not using combined dressings, and there is evidence that interactions do occur which increase yields above those obtainable with equal quantities of nutrients from either source alone. These interactions can be explained by:

i. improvements in soil structure caused by the compost leading to better moisture retention characteristics which improve the efficiency of use of mineral fertilizers;

ii. increases in the availability of phosphorus in the soil and from mineral fertilizers caused by the application of compost;

iii. more immediate availability of nutrients from mineral fertilizer sources than from composts and longer term release of nutrients from composts;

iv. reduced leaching of nutrients;

v. increased activity of soil microbes and animals.

The benefits obtained vary from area to area, from season to season and with the crops grown. They tend to be less with larger applications of both fertilizer and compost. In general the following guidelines can be given:

a. mixtures of mineral fertilizers and composts prevent deficiency diseases;

b. mineral fertilizers and composts used together preserve or improve soil fertility and complement each other;

c. best results are obtained when at least 30 percent of the nitrogen is supplied by each source;

d. the most appropriate combination of type and percentage of mineral fertilizer to use can be found by local trials.

5.12 FISH FARMING

The production of fish under intensive conditions in ponds is a well established practice in some parts of the tropics, particularly in Asia and, to a lesser extent, in Africa. Organic manures and wastes can be added to the ponds to provide a direct source of feed for the fish or to provide nutrients for the growth of plants or organisms upon which the fish can feed. When the organic manure or fertilizer is added the concentrations of bacteria and plankton in the pond increase. Fish such as the tilapia species feed both on plankton and on detritus and are very responsive to the use of organic manures and fertilizers; their yields can be increased considerably. Compost feeding of fish ponds is also successful with other fish species such as those of the carp family (Cyprinidae).

The main advantage of using compost instead of the original raw wastes is that the majority of the 'oxygen demand' of the wastes has been met during the composting process. Thus a greater amount of compost than original wastes can be added to the fish pond without adversely affecting the oxygen levels in the pond.

Banerjee and Srinivasan (1983) investigated the possibility of using composted urban refuse and primary sewage sludge as a fish pond manure. The compost was produced from the compostable fraction of Calcutta refuse mixed with sewage sludge from a primary settling tank. The compost was added at 200 mg/litre to 400 litre capacity pools and the productivity measured. Plankton, the main natural fish food organism, was used as an indication of the productivity. The compost addition caused a level of about 2100 units per litre of phytoplankton, ten times the level present without the addition of compost. The experiment showed that it was better to use compost than conventional farmyard manure and that the amount of compost added could be increased safely to 600 mg/litre without adversely affecting the oxygen levels in the ponds. It is suggested that aquaculture responds better to organic manures and fertilizers than does agriculture.

Edwards *et al*. (1983) reported on the addition of compost, produced from night soil and water hyacinths in above-ground heaps, to a series of 200 m^2 earth walled ponds containing tilapia fish. The latter consumed the compost directly and produced significant fish yields at a mean conversion rate of 7.4 units of food consumed per unit of body weight gained, compared with 4.6 for cereals, assuming comparable food moisture contents.

CHAPTER 6

ENVIRONMENTAL ASPECTS

6.1 GENERAL

Some of the materials suitable for composting can be the source of public health problems. Most organic wastes from human and animal communities will be contaminated to some extent with pathogenic organisms which can cause infections in man, animals and plants. Examples of these wastes are night soil, sewage sludge, animal manures and, to a smaller extent, refuse. Crop wastes may also carry some plant pathogens. Easily degraded materials such as food scraps, together with night soil and manures, are also attractive to flies and vermin which can spread disease. In addition to public health problems, there may be difficulties from unpleasant smells and the spread of weed seeds when using the compost product. Careful attention to meeting the process requirements of composting can very significantly reduce the environmental problems caused by waste materials. It is important that the waste be brought into a biologically acceptable condition so that litte or no hazard will then be presented, either to those handling the compost or to those consuming crops grown using the compost. The objective should be to bring the compost product to a biological state comparable to that of the soil into which it will be incorporated.

6.2 PATHOGENS

Disease-causing pathogenic organisms may belong to any of the main types of micro-organisms - bacteria, actinomycetes, fungi, viruses, rickettsiae and protozoa. They also include macro-fauna, helminths/nematodes or intestinal worms. Most of these pathogenic organisms prefer temperatures below 42°C because they normally live at the body temperatures of man and animals or the air temperatures experienced by plants. Most will die if exposed for sufficient time to conditions more severe than their usual environment.

In a composting process there are several factors which act against the survival of pathogenic organisms. These are:

i. the high temperature reached;

ii. the time that the process takes;

iii. the release of ammonia during the process which has a direct effect on some organisms and alters the waste from acid to alkaline conditions;

iv. the intense competition for food, particularly during the cooling-down stage;

v. competition and attacks between the micro-organisms at the start of the maturing stage, together with the release of some antibiotic substances.

The temperature-time relationship is the most significant factor in causing the death of pathogens. Above a certain temperature level, a higher temperature for a short period of time or a lower temperature for a longer period can have the same effect. An example of this is shown by the effect of temperature upon the organism _Salmonella typhi_, which causes typhoid fever. It does not grow above 46°C; at 55°C it dies in 30 minutes but at 60°C it dies in 20 minutes. Figure 37 shows the temperatures which are required to kill most organisms in soil, using heat for 30 minutes under moist conditions. Table 20 lists many of the common pathogens which cause diseases in humans and domestic animals, with conditions of temperature and time under which they will be killed. This shows that they are readily destroyed at temperatures of 55 to 60°C for periods of a few minutes to a few hours under the moist conditions used in composting. The ova of intestinal worms are similarly destroyed under such conditions. However, although most disease and parasitic organisms are killed, a few spore-forming bacteria, if present, can survive temperatures in excess of 100°C. Examples are the aerobic spore-forming Bacillus anthracis, the organism that causes anthrax, and the anaerobic spore-forming Clostridium species, which cause tetanus, botulism and gas gangrene. These organisms are likely to survive the composting process. This is not a serious disadvantage as these organisms are normally widespread in the soil.

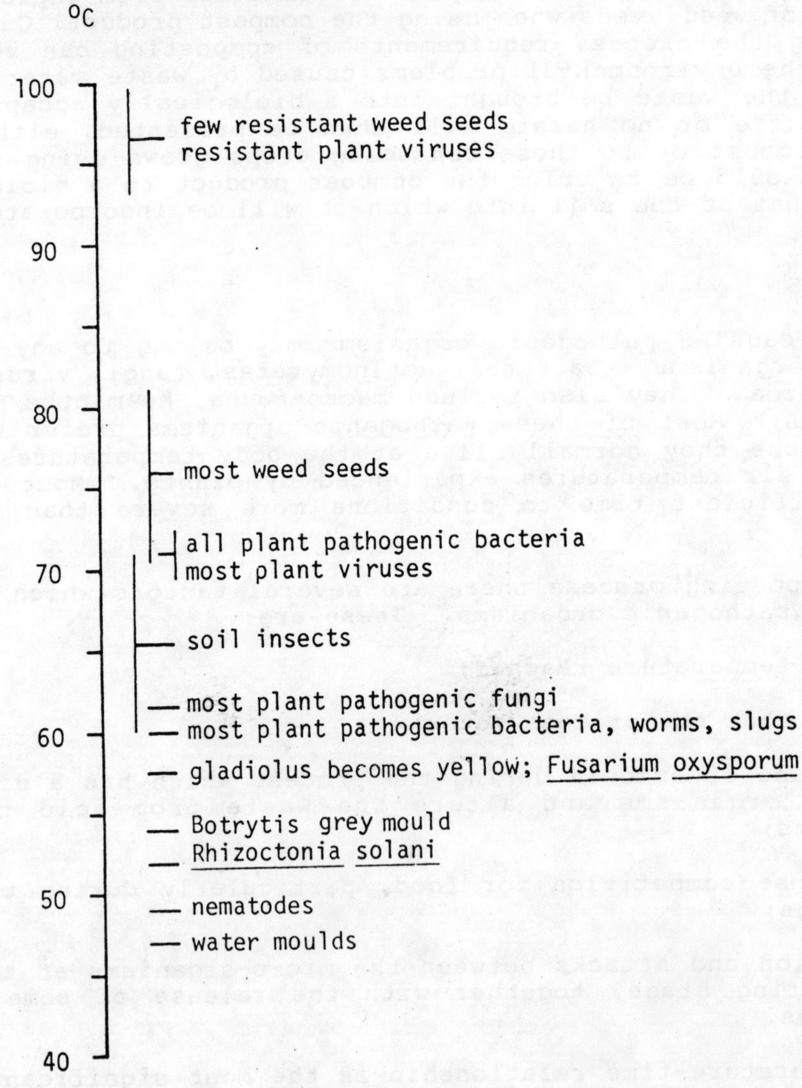

Figure 37 Soil sterilization temperatures

Table 20 LETHAL TEMPERATURE-TIME CONDITIONS FOR SOME COMMON PATHOGENS AND PARASITES

Disease	Organism	Lethal conditions (moist heat)
Anthrax	Bacillus anthracis	Over 100°C
Brucellosis or Contagious abortion	Brucella species	10 mins - 60°C
Cholera	Vibrio cholerae	15 mins - 55°C
Diphtheria	Corynebacterium diphtheriae	10 mins - 58°C
Dysentry	Shigella species	60 mins - 55°C
Food poisoning	Salmonella species	20 mins - 60°C
Leptospirosis (Weil's disease)	Leptospira species	10 mins - 50°C
Plague	Pasteurella pestis	5 mins - 55°C
Staphylococcal infections	Staphylococci	30 mins - 60°C
Streptococcal infections	Streptococci	30 mins - 55°C
Tuberculosis	Mycobacterium tuberculosis	20 mins - 60°C
Typhoid fever	Salmonella typhi	20 mins - 60°C
Vibriosis	Vibrio fetus	5 mins - 56°C
Intestinal worms		
Round worm	Ascaris lumbricoides	60 mins - 55°C
Tape worm	Taenia saginata	A few minutes - 55°C
Hookworm	Ancylostoma duodenale and Necator americanus	1 min - 55°C

It is impossible to be certain that a particular composting process will produce a product which is entirely free of pathogens. However, if composting is correctly carried out so that temperatures of 55 to 60°C are reached for a few days, an acceptable hygienic product should be obtained. The important question is whether all of the material in the composting heap has reached such high temperatures. In a windrow heap there can be a large variation in temperature and air availability, as shown in Figure 20. There is a cool outer layer, cool pockets at the lower edges, possibly an anaerobic region at the centre bottom and a high temperature region just above the middle. The width and height of the heap are normally limited by the need to allow air to move naturally into the heap. If the heap is forced air blown with a fan and aeration pipes at the bottom of the heap, there should not be an anaerobic region but there will still be a higher temperature region above the middle. If air is sucked down through the heap, the hottest region will be at the centre bottom.

If a natural air flow heap is being used, it will often be turned to mix and provide air; it is important when turning that the material from the cooler parts of the heap should then be placed in the central hotter region of the newly formed heap. Each time the heap is turned, material from the edges should be placed in the middle so that as much as possible of the material reaches the highest available temperature. Another means of raising the temperature of more of the material is to cover the heap with an insulating layer. This layer should be at least 25 mm thick and materials such as finished compost product, straw or mud can be used. The layer will reduce the loss of heat from the heap and will become the cooler outer region so that more of the interior fresh material will reach the higher temperatures.

Totally enclosed, mechanically agitated composting systems more easily subject most of the material to the higher temperatures. In an agitated enclosed reactor the heat is spread fairly evenly throughout the mass; there should be no major over-heated or under-cooled regions. However, such systems are expensive to install and operate.

A number of tests of pathogen survival in composting systems have been reported. It is evident that, in a composting mass which is well agitated, or is turned fairly often, or is well insulated so that all the material is exposed to the higher temperatures for a period of several hours, all the common pathogenic organisms investigated are significantly destroyed. Composting heaps are not the natural environment for pathogens and they will tend to be eliminated in such heaps. In those situations in which highly resistant spore-forming pathogens may be found, the organisms are very likely to be present in the soil and the composting process will not offer a degree of sterilization of those organisms better than the soil to which the product is to be applied.

Some recent publications have extensively reviewed the literature on the presence and survival of pathogens in composting systems (Finstein *et al*. 1982; Higgins 1983; Golueke 1983).

6.3 FLIES AND VERMIN

Another important aspect of some of the materials that can be used in composting is their attractiveness to flies and vermin. It is manure-type materials, of human or animal origin, that present the greatest problem. The fly larvae in composting material can originate from eggs laid in the material at its origin or from eggs laid during handling at the composting site. A typical life cycle of a fly is as follows: egg, 1-2 days; larva, 3-5 days; pupa, 3-5 days; young fly, 7-11 days; egg laying by new fly, 10-14 days. The objective should be to interrupt this cycle and prevent development of adult flies. Temperature is an important factor in interrupting the cycle. As the material being composted passes through its temperature peak and becomes stabilized, it is less attractive to flies and vermin.

To avoid the problem, the recommendation is the same as for causing a reduction in pathogens, that is to subject as much of the material as possible to higher temperatures. Fly larvae are unlikely to survive in regions where the temperature is above about 55°C; they are likely to be found in the cooler outer layers of a heap. By turning the heap and placing the outer material in the hot central region many of the larvae will be destroyed; satisfactory fly control is possible by turning. The insulating layer approach is also effective. A covering layer of material such as product compost, straw or mud is not attractive to flies and vermin and raises the temperature of the material inside the heap. Acharya (1950) reported successful control of flies in composting mixtures of municipal refuse and night soil by placing an insulating layer of refuse covered with a 50 mm thick layer of dirt over the heap.

Fly control is not normally a problem in totally enclosed, mechanically agitated composting systems because virtually all of the material reaches temperatures of 55°C and above, and the material is stabilized to a point where it is no longer attractive to flies when it is removed from the reactors.

Flies and vermin can present a serious public health problem. Providing that the waste materials that cause this problem are properly composted, reach temperatures of about 55°C and are stabilized, the problem is effectively controlled.

As mentioned in Section 5.1, compost heaps can provide breeding sites for the rhinoceros beetle in Western Samoa if the compost is not used within three months. In many tropical countries termites can use the compost heaps for nest building unless they are prevented.

6.4 ODOURS

Breakdown of organic matter by aerobic oxidation produces no objectionable odours. If odours are present, either the process has less air available than the amount required for the process to be entirely aerobic, or there are materials present, other than from the oxidation reaction, which have an odour. The breakdown of material in the absence of air, anaerobic breakdown, usually results in the release of the unpleasant smells of hydrogen sulphide and other sulphur compounds. As shown in Figure 20, it is possible, in a compost heap using a natural air flow through the material, for an anaerobic region to be present. This can be the source of unpleasant smells. It is important therefore to supply sufficient air, not only for the composting process breakdown but also to prevent an odour problem. The use of an insulating layer for reducing heat loss in order to raise the temperature for pathogen and fly control can also help reduce smells. If compost product is used as the insulating layer, the compost will act as an odour filter and absorb some of the compounds responsible for the unpleasant smell.

Once the composting material has passed its peak temperature and reached the point of stability, most of the sulphur and nitrogen which can cause odours have then become bound up in the bodies of new micro-organisms. Thereafter odours should not be generated.

6.5 WEEDS AND SEEDS

Annual weeds only propagate from seeds, whereas biennial and perennial weeds can propagate both from seeds and from fragments of roots. A properly made and controlled compost heap should kill off most weed roots and seeds present. However, a badly made heap which does not heat up adequately may leave such material in a viable state; as a result the product compost could give rise to a profusion of weed seedlings when used.

The vast majority of seeds should be killed at temperatures of about 60°C, held over a period of perhaps 3 days. However, a few very resistant seeds with hard coats will require much higher temperatures and may possibly survive composting. Figure 37 illustrates the temperatures required to kill seeds, insects and plant pathogens when heated for only 30 minutes under moist conditions. As with the destruction of pathogenic organisms mentioned earlier, a temperature-time relationship exists for killing seeds. Hence temperatures about 10-15°C lower than those shown in Figure 37 should suffice in a compost heap in which a peak temperature of over 60°C is maintained for a period of several days. This should be possible with heaps containing about 2 m^3 or 1 tonne of wastes. With smaller heaps the high temperatures are only achieved in a small section in the upper centre of the heap; most of the mass is at the cooler margins. In such cases, the cool sides need turning in the middle to obtain an effective kill of weeds and seeds.

In the case of hardy, persistent biennial and perennial weeds it is better not to put the roots into a compost heap. It is far safer to burn them, thereby ensuring their destruction.

6.6 HEAVY METAL CONTAMINATION

In heavily industrialized areas, where urban wastes used for compost production may contain significant concentrations of toxic heavy metals, it is necessary to prevent long-term build-up of such metals in the soil. Of the metals that may be present those which are most likely to be toxic to plants are copper, nickel, zinc and possibly boron. Other metals such as cadmium, lead or mercury may also be present but their concentrations are not normally high enough to have an adverse effect upon plant growth. The problem is, however, that they may accumulate in plant tissue to give concentrations which might be harmful to the health of humans or animals eating the crop. There is particular concern regarding cadmium as it moves relatively easily from plant roots to the above-ground parts of a plant.

Although research has been carried out on the effect of metals in the soil, there is no clear and simple understanding of the problem. It is known that the availability of metals in soils to plants is very dependent on the soil reaction (pH). In soils which are acidic most toxic metals are quite readily available; in soils which are naturally alkaline or have been made more alkaline by improved drainage or the addition of lime in some form, trace metal availability is often greatly reduced. Other factors are also important and these include the organic matter content of the soil, the quality of the organic matter, the presence of other elements, soil temperature and the plant species growing.

Field trials have been reported by Gray and Biddlestone (1980) in which various UK urban composts were added to alkaline clay soil and acid sandy soil. Trace metal analyses of the composts and the soils showed that the urban waste composts contained appreciable levels of some trace metals such as copper, lead and zinc and that these can significantly increase the total and extractable (available) levels in soils to which they are applied. Crop analyses, however, revealed that the trace metals were not taken up in the same proportion as the increase in the soil. Even with heavy dressings of compost the crops did not exceed existing statutory or recommended limits for metals in foodstuffs.

The real problem is that once these metals have been applied to the soil they cannot easily be removed, apart from water-soluble boron. Although they may give rise to little, if any, toxicity if small dressings are applied regularly in good husbandry, there is the possibility that, if the soil pH and organic matter content are allowed to fall markedly, increased metal availability to plants will occur. The long-term objective should be to prevent such metals entering organic waste materials which are likely to be recycled back to the soil. It is a problem normally only encountered in industrialized areas generating metal-contaminated trade waste; appropriate monitoring of the concentration of metals in these wastes is then recommended before they are used for producing composts. In most less developed countries, the urban waste from cities and towns is fairly effectively scavenged and most solid metal objects removed for recycling to uses other than agriculture.

6.7 CONTAMINATION OF DRINKING WATER

The widespread use of water-soluble mineral fertilizers in temperate zones has resulted in the washing into water sources of significant quantities of nitrate (NO_3) nitrogen. This poses health hazards if the water is used for human consumption. An NO_3 level in excess of 45 parts per million (ppm) can cause harm to children.

Studies have been carried out in India to see if the use of soluble mineral fertilizers led to water pollution in situations involving high application rates and intensive irrigation in light soils. Results showed that the levels of NO_3 were much higher in wells close to the villages than in the cultivated areas. Improperly handled animal wastes were thought to be the contributing factor to these high levels, many of which exceeded 45 ppm. In cropped areas where mineral and organic manures were used, there was a balance between N, P and K and growing crop roots which absorbed most of the nutrients, thus minimizing losses to groundwater. These results show the importance of controlled methods of handling animal manures and emphasize the relevance of composting.

As a further safety measure, it is recommended that no compost pit be set up within 25 m of a drinking water source (Figure 38). This should prevent any liquid percolating from the compost heap into the water supply, especially during the rainy season.

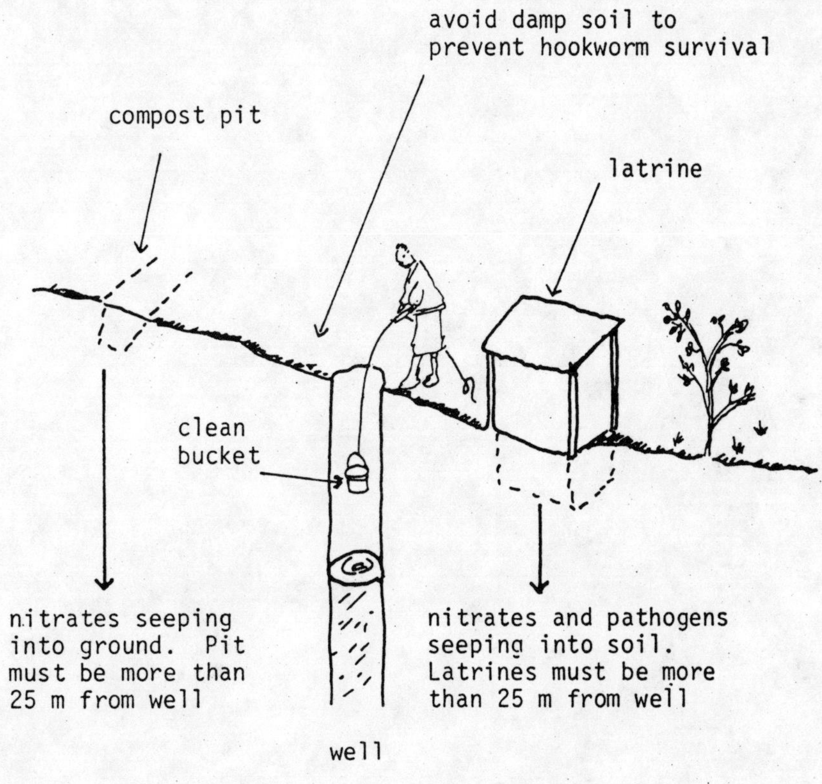

Figure 38 Water pollution precautions

CHAPTER 7

ECONOMIC AND SOCIAL ASPECTS

7.1 GENERAL

Most farmers in the developed world can purchase whatever quantity of mineral fertilizer they wish to use. Because of the high cost of labour and mechanization they may not be able to handle large quantities of organic materials. Increasingly, farm management in the West is becoming an annual economic consideration controlled by production quotas and price support systems. Large surpluses stored in national and international warehouses mean that maximization of production is not always profitable and that each year's farm production plan may be quite independent of the preceding or following year's plan. These factors explain the decrease in the practice of farm-scale composting in the period from 1950-1970. However, there has recently been renewed interest in the West in composting, mainly for the following reasons:

i. the economic pressure of the oil crisis and diminishing sources of raw materials for manufacturing mineral fertilizers;

ii. problems of pollution caused by the dumping of untreated wastes. In spite of the widespread occurrence of such problems, it is increasingly apparent that preventative methods are unlikely to be adopted without legislative action;

iii. difficulties in alternative methods of farm waste disposal such as direct incorporation of straw into the soil or else burning it;

iv. concern about deterioration of soil structure if levels of soil organic matter fall;

v. increasing interest in, and economic premiums on, organically grown health foods.

The economic situation differs greatly in the developing world, a major portion of which falls in the tropical belt. In many areas the continuance of farming from year to year is dependent on soil conservation and the success of production in a season is dependent on soil moisture management. National food surpluses are an exception and for many farmers production is directly linked to their own food requirements for survival.

Labour costs in the less developed countries are much lower than in the West. Labour is much more readily available and in cases where composting would utilize family labour, no cash outlay would be required. Collecting, composting and spreading organic wastes would create work opportunities for under-employed labour. Raw materials for composting are not usually subject to artificial shortages due to local or international market fluctuations. Unlike oil or raw materials for mineral fertilizers, they are locally renewable. It is true that increasing emphasis on composting might well lead to a shortage of some organic wastes, but this could be largely offset by:

a. modification of the farming system to produce organic materials in greater quantities. This can be done without reducing yields, e.g. by planting leguminous trees and hedges, and by better control of grazing to reduce dependence of livestock on straw products;

b. use of organic wastes which are at present employed for other purposes, e.g. cattle dung is often used for fuel, sugarcane trash is normally burnt on the field;

c. use of organic wastes which are at present discarded, e.g. night soil or sewage sludge, forest litter, paddy husk.

Farm transport is not only cheaper than in the West but, being mainly animal-drawn, relies less on oil. In many cases, the transport of materials would add little to total farm transport costs as it could be done when existing vehicles are not required for other work.

The farmer may face the dilemma of not having sufficient inorganic fertilizer or sufficient irrigation for all his land. The likelihood is that all available mineral fertilizer will be used on irrigated land and none on rainfed, un-irrigated land. He may also be tempted to use all his cash for mineral fertilizer and ignore such possibilities as better seed, pesticides and improved equipment. Composting would enable him to increase the input of nutrients, to spread the usage of mineral fertilizer over all his land, and to save cash for other improvements.

In dryland farming conditions in the semi-arid tropics, there are many tracts of soil which would support much larger yields of crops given better seed, small dosages of mineral fertilizers and better soil and water management. There is evidence to show that on poorer soils, if the level of production is increased, then the soil becomes depleted of trace elements; deficiencies appear which are not seen under the traditional farming pattern. Such deficiencies can be prevented by the use of organic manures and composts.

It must be emphasized, however, that composting is not being advocated as a solution for all problems. There are situations where sufficient organic raw materials would not be available or where the making of large quantities of compost would not fit easily into the farming system. The advantage of composting, apart from its effect on soil structure and moisture retention, is that in many situations it is efficient in economic terms and has definite social benefits, some of which are illustrated in Figure 39.

7.2 ECONOMIC CONSIDERATIONS

The decision on whether or not to make compost will depend largely on the cost and availability of labour, organic raw materials and mineral fertilizer in the farmer's locality (Figure 39). These factors can vary from year to year and so the relative cost of plant nutrients between organic and mineral sources will also vary.

In making comparisons between composts and mineral fertilizers on a nutrient basis several points must be borne in mind:

i. composts vary widely in composition depending on the raw materials used in their preparation. It is unlikely that the total amount of any nutrient will be greater than that present in the original raw material;

ii. composts usually provide all the major plant nutrients, nitrogen (N), phosphorus (P), potassium (K), and a wide range of micro-nutrients, whereas mineral fertilizers do not;

iii. nutrients are normally released more slowly from composts than from very water-soluble mineral fertilizers. Consequently the effects of compost can last for more than one season; loss of nutrients from the soil by leaching during the growing season is reduced by the use of compost.

Details of the percentages of N, P and K in some of the more common mineral fertilizers and in a number of composts are presented in Tables 21 and 15. In Table 21 phosphorus is given as P and not P_2O_5, and potassium as K and not K_2O.

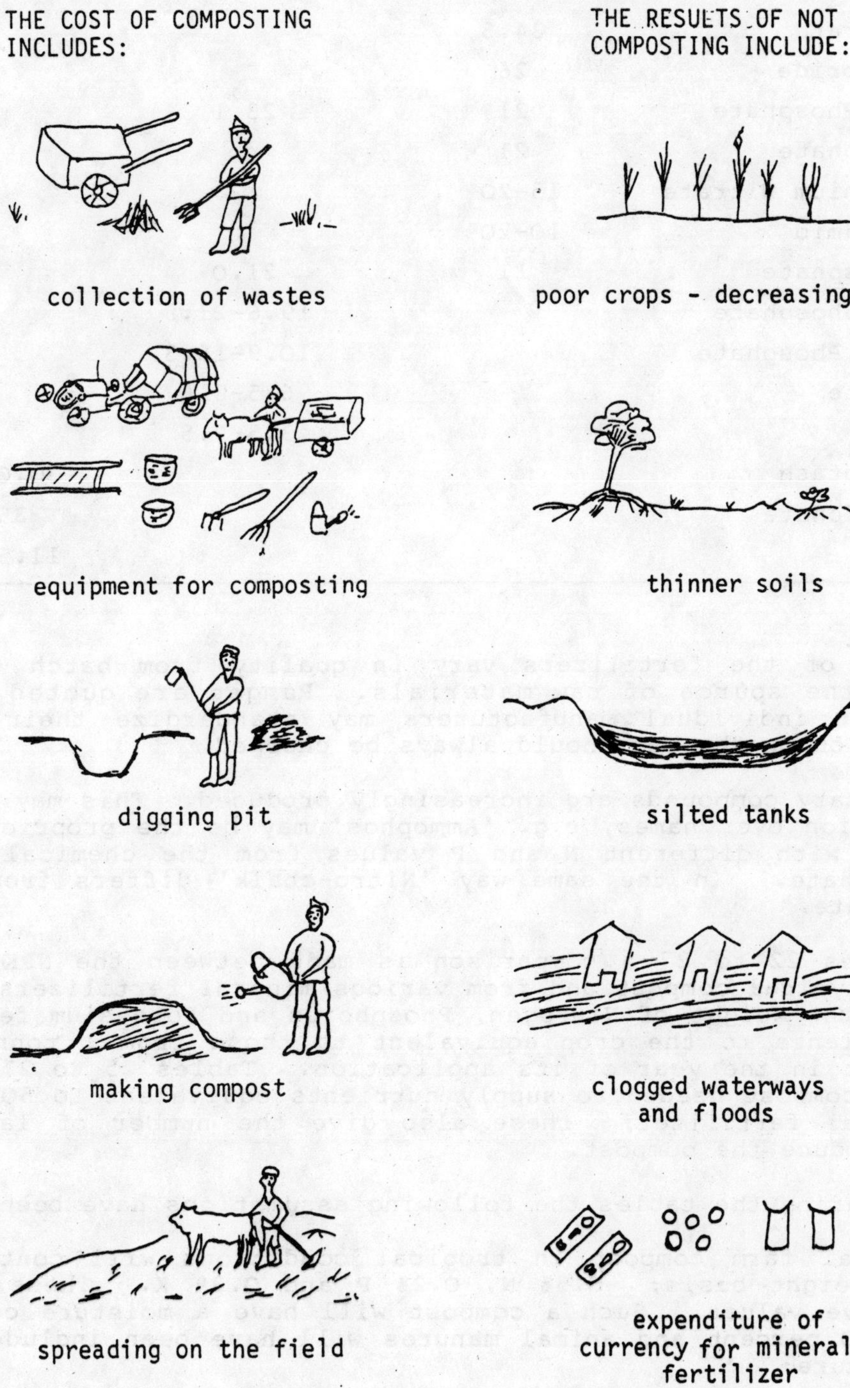

Figure 39 Costs of making and not making compost

Table 21 COMPOSITION OF MINERAL FERTILIZERS

Fertilizer	Percent N nitrogen	Percent P phosphorus	Percent K potassium
Urea	46		
Ammonium Nitrate	34.5		
Ammonium Chloride	26		
Di-ammonium Phosphate	21	23.1	
Ammonium Sulphate	21		
Calcium Ammonium Nitrate	15-20		
Calcium Cyanamid	10-20		
Ammonium Phosphate	11	21.0	
Triple Superphosphate		19.6-21.8	
Mineral Rock Phosphate		10.9-15.3	
Superphosphate		6.5-8.6	
Basic Slag		3.5-9.5	
Muriate of Potash			30.0-51.5
Potassium Sulphate			35-39
Kainit			11.5-16.5

Several of the fertilizers vary in quality from batch to batch, depending on the source of raw materials. Ranges are quoted for such fertilizers but individual manufacturers may standardize their product. The specification on the bag should always be checked.

Proprietary compounds are increasingly produced. This may give rise to some confusion over names, e.g. 'Ammophos' may be the proprietary name for a complex with different N and P values from the chemical compound ammonium phosphate. In the same way 'Nitro-chalk' differs from calcium ammonioum nitrate.

In Tables 22 to 27 a comparison is made between the NPK nutrient supply from a typical compost and from various mineral fertilizers. Tables 22 to 24 list the weight of Nitrogen, Phosphorus and Potassium fertilizers supplying nutrients to the crop equivalent to those from 1 tonne of the typical compost in the year of its application. Tables 25 to 27 show the tonnes of the compost needed to supply nutrients equivalent to 50 kg of N, P and K mineral fertilizer; these also give the number of labour-days required to produce the compost.

In preparing the tables the following assumptions have been made:

a. a typical farm compost in tropical conditions will contain on a fresh weight basis: 0.5% N, 0.2% P and 0.3% K. These are conservative values. Such a compost will have a moisture content of about 50 percent and animal manures will have been included in its manufacture;

b. all the nutrients in mineral fertilizers will be available to the crop in the year of application;

c. the following percentages of nutrients in the compost will become available to the crop in the year of application:

Nitrogen - 25%; Phosphorus - 100%; Potassium - 80%.

No account has been taken of the availability in succeeding years;

d. to prepare and spread one tonne of compost will require 3 labour-days.

The assumption of 3 labour-days for the production and spreading of one tonne of compost is based on recent experience at the Medak Agricultural Centre, India, using the equipment outlined in this manual and turning twice. It is similar to the figure achieved by Howard at Indore in India in the nineteen-twenties, but higher than the figure of 2 labour-days per tonne reported from an estate at Chipoli in Rhodesia in the late nineteen-thirties. The actual labour requirement will vary with the type of wastes, the transporting distance and the composting process used.

It should be noted that much of the work involved in composting can be carried out by female labour. Where wage rates are low, Tables 25 to 27 indicate that compost can provide a cheap alternative source of nutrients to many of the less concentrated single nutrient fertilizers. When considering the supply of the three major nutrients, compost may compete economically on a nutrient basis with the more concentrated fertilizers.

For example:

Application		Nutrients supplied to crop in kg		
		Nitrogen	Phosphorus	Potassium
1.	Mineral fertilizers			
	- Urea, 27.3 kg	12.5	-	-
	- Triple Superphosphate 100.0 kg	-	20.5	-
	- Muriate of Potash, 45.9 kg	-	-	23.6
	Total fertilizer 173.2 kg	12.5	20.5	23.6
2.	10 tonnes typical compost	12.5	20.0	24.0

The labour cost of 10 tonnes of compost is 30 labour days. In many areas this will be less costly in cash terms than the combination of mineral fertilizers shown as being equivalent in nutrients supplied. These calculations take no account of the beneficial effect of compost on trace element supply, soil structure and water holding capacity.

A major benefit of composting is that it reduces the volume and weight of wastes and thus eases transport problems. Accordingly, it is a good idea to do the composting as close as possible to the source of the main organic raw materials, although remembering that water will also be needed in large quantities, Figure 40. It is widely accepted that the efficiency of repetitive manual transport processes will be markedly increased by the use of wheelbarrows if individual trip distances are more than 15 metres. Processes which involve repeated turning of othe heap obviously require more labour than single-turn methods. Frequent turning decreases the time required from building the heaps to use of the product compost and if time is limited the additional labour cost may be inevitable.

Table 22 NUTRIENT COMPARISONS OF COMPOST WITH MINERAL NITROGEN FERTILIZERS

Fertilizer	Weight of fertilizer in kg supplying N to crops equivalent to 1 tonne of compost in year of application
Urea	2.7
Ammonium Nitrate	3.6
28-28-O Complex	4.5
Ammonium Chloride	4.8
Ammonium Sulphate	6.0
Di-ammonium Phosphate	6.0
Calcium Ammonium Nitrate	6.3-8.5
15-15-15 Complex	8.3
Ammonium Phosphate	11.4
Calcium Cyanamid	6.3-12.5

Table 23 NUTRIENT COMPARISONS OF COMPOST WITH MINERAL PHOSPHORUS FERTILIZERS

Fertilizer	Weight of fertilizer in kg supplying P to crops equivalent to 1 tonne of compost in year of application
28-28-O Complex	7.0
Di-ammonium Phosphate	8.7
Ammonium Phosphate	9.5
Triple Superphosphate	9.0-10.0
Mineral Rock Phosphate	13.1-18.3
15-15-15 Complex	13.3
Superphosphate	23-31
Basic Slag	21-57

Table 24 NUTRIENT COMPARISONS OF COMPOST WITH MINERAL POTASSIUM FERTILIZERS

Fertilizer	Weight of fertilizer in kg supplying K to crops equivalent to 1 tonne of compost in year of application
Muriate of Potash	4.7-6.2
Potassium Sulphate	6.2-6.9
15-15-15 Complex	16.0
Kainit	14.5-20.9

Table 25

NITROGEN EQUIVALENT
(tonnes of compost supplying equivalent amount of N to crops in year of application as 50 kg of mineral fertilizer)

Fertilizer	Tonnes compost equivalent to 50 kg fertilizer	Labour-days to produce compost
Urea	18.4	55
Ammonium Nitrate	13.8	41
28-28-0 Complex	11.2	34
Ammonium Chloride	10.4	31
Ammonium Sulphate	8.4	25
Di-ammonium Phosphate	8.4	25
Calcium Ammonium Nitrate	6.0-8.0	18-24
15-15-15 Complex	6.0	18
Ammonium Phosphate	4.4	13
Calcium Cyanamid	4.0-8.0	12-24

Table 26

PHOSPHORUS EQUIVALENT
(tonnes of compost supplying equivalent amount of P to crops in year of application as 50 kg of mineral fertilizer)

Fertilizer	Tonnes compost equivalent to 50 kg fertilizer	Labour-days to produce compost
28-28-0 Complex	7.0	21
Di-ammonium Phosphate	5.8	17
Ammonium Phosphate	5.3	16
Triple Superphosphate	4.9-5.5	15-17
Mineral Rock Phosphate	2.7-3.8	8-11
15-15-15 Complex	3.8	11
Superphosphate	1.6-2.2	5-7
Basic Slag	0.9-2.9	3-9

Table 27

POTASSIUM EQUIVALENT
(tonnes of compost supplying equivalent amount of K to crops in year of application as 50 kg of mineral fertilizer)

Fertilizer	Tonnes compost equivalent to 50 kg fertilizer	Labour-days to produce compost
Muriate of Potash	8.0-10.8	24-32
Potassium Sulphate	7.3-8.1	22-24
15-15-15 Complex	3.1	9
Kainit	2.4-3.4	7-10

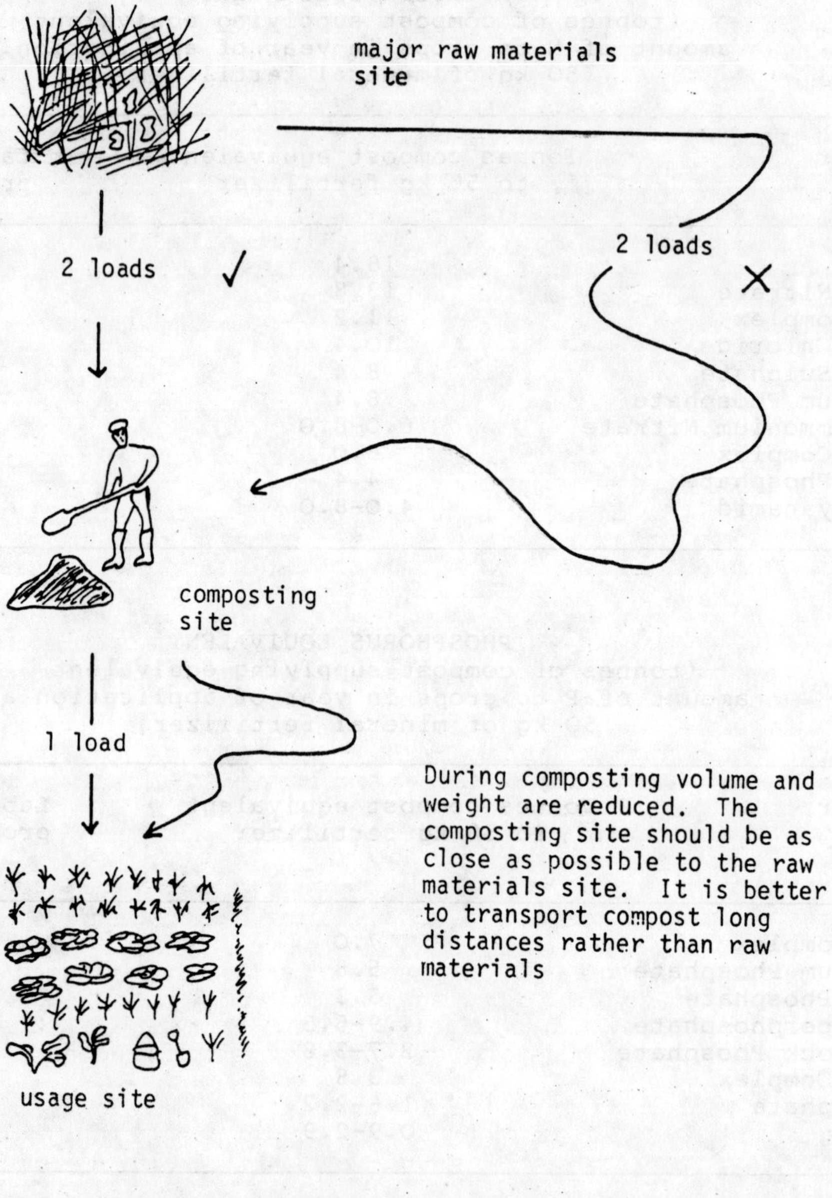

Figure 40 Efficient use of transport

There is little sale value for most of the wastes used for composting. The main exception is for dung which may have a local marketable value. This is not a serious problem if the farmer is already spreading some dung on his land. It could become a problem in areas where biogas plants are becoming popular as there may then be an increased price for dung. Small farmers may be tempted to sell all their stored animal manure for the initial charging of a biogas plant and hence will not be able to produce good quality compost.

The disposal of animal wastes is an inescapable daily routine in many parts of the tropics. Often dung and spoiled fodder are dumped haphazardly in pits close to the animal sheds or on the outskirts of the village. Much of the nitrogen is then lost as ammonia. As the material dries out quickly, very little decomposition takes place, unless it is attacked by termites, with the result that it may be lost to the agricultural cycle. Systematic composting will make better use of the labour at present used in the disposal process, conserve vital nutrients and result in a more effective product. The cost in terms of land used is unlikely to be a serious problem as it is probable that improved composting can be carried out on land already employed for waste treatment. If additional land is required, it will almost certainly be possible to find non-cultivable land with a low potential for alternative use.

7.3 SOCIAL CONSIDERATIONS

Composting could be an integrated part of the total process of waste disposal in a community and especially night soil disposal. In most parts of the tropics ecological and public health considerations indicate the necessity for widespread composting. Reference to Chapter 6 will provide details of the way in which composting can minimize the spread of pathogenic diseases, control odour, fly and vermin problems and prevent pollution of drinking water supplies. Many village committees spend considerable sums of money on night soil collection without safety controls on pathogens and without achieving a useful product. Several governments spend much money on 'food for work programmes' with little long-term benefit. These investments might render a better return if they were directed to introducing composting, thereby easing waste disposal, providing a saleable product and some employment. The use of compost would also help combat soil erosion, thus reducing the necessity to clear silt-filled dams and helping to prevent flooding.

The advent of mineral fertilizers and pesticides has changed farming in many areas from a subsistence to a commercial basis. There is evidence that the efficiency of use of these inputs is low and that the improvements which result are less than 60 percent of those which should be achievable. There can be little doubt that the use of these inputs will be much more efficient if linked to a progressive policy of soil management involving the complementary use of mineral and organic fertilizers.

The full social benefits of composting will only be achieved if changes are also made in fodder and fuel practices. Thus biogas production will obviate the need to burn dung and will yield a compostable sludge product. Improved fodder production schemes will release high C/N ratio wastes for composting; however, such schemes may require changes in community grazing patterns.

There can be little doubt that there are many important social benefits to be gained by the increased use of composting. Equally it is apparent that deeply entrenched social practices and prejudices are major obstacles to the increased use of composting. These are mostly connected with night soil handling and are unlikely to be overcome without appropriate agricultural and public health education. Hence coordination of the work of various local government departments is necessary.

CHAPTER 8

EDUCATION AND TRAINING OF FARMERS AND EXTENSION STAFF

8.1 GENERAL

Several difficulties must be overcome if the practice of composting by farmers is to become more widespread in tropical agriculture. These are:

i. transport problems;

ii. shortage of raw materials;

iii. alternative use of wastes, such as dung burning;

iv. social distastes associated with night soil handling;

v. possible health hazards;

vi. inadequate research services;

vii. lack of knowledge on the part of farmers about the limited effects of mineral fertilizers and of the importance of organic matter in maintaining soil fertility;

viii. deficiencies in extension services.

Suggestions for dealing with all these problems are given in this manual.

Difficulties (iii), (iv) and (v) are connected with long established social practices; they are not directly agricultural but cannot be ignored. Farmers should be educated to see how agricultural progress may only be achieved through social change. The remaining problems are less difficult to solve, given an integrated approach by farmers, extension staff and reserach workers. One promising method of achieving this integration is the 'Training and Visit' system of agricultural extension (Benor and Baxter 1984). This system has been designed for Government Extension Services but can readily be adapted to the circumstances of non-government agencies.

8.2 REQUIREMENTS OF EXTENSION SERVICES

Whether a government or a non-government agricultural extension system is used it must meet four requirements:

i. initial improvements should aim at increasing the efficiency of the existing agricultural system. Low cost improvements which yield good results should be tried and will lay the foundations for later and more costly changes;

ii. the extension worker should receive regular training from his supervisors and from subject matter specialists. Each training session will equip him for his next round of village visits and give him an opportunity for reporting back on his last round;

iii. the worker should make regular predetermined visits to his village, either on a weekly or fortnightly basis. He should have a clear idea of what tasks he has to perform. The farmers will learn quickly that they have access to skills beyond those of the extension worker. If necessary, specialist staff can be brought to the village to deal with a particular problem;

iv. key farmers should be encouraged to test new methods on small areas of their own land. In this way local knowledge can be built up, new techniques demonstrated, risk minimized and practical problems relayed back to researchers.

The way in which a composting project can fit into such an extension system will now be considered.

8.3 TIMING OF THE EXTENSION PROGRAMME

Planning of the extension programme should commence with consideration of the time required for individual aspects, so that the whole can be fitted into the farming year. Possible schedules are given in Table 28. The earliest planting date for main season crops is a fixed point in time. Working back from this date it is possible to calculate the latest date for building a compost heap which will allow adequate time for processing and spreading the product. The minimum period for these operations is 14 weeks using the Indore method which has a relatively high labour demand and hence costs more than alternatives such as the Bangalore process. If organic wastes with a high C/N ratio are used, together with a single turning, then a period of 30 weeks from making the compost heaps to crop planting may be needed.

Table 28 SCHEDULE FOR COMPOST EDUCATION, MAKING AND USE

Stage	Stage time, weeks		Time from start, weeks	
	Minimum	Maximum	Minimum	Maximum
Planning and initial extension worker training	4	4	4	4
Soil fertility teaching to farmer	1	8	5	12
Collection of wastes, (in addition to daily collections)	3	6	8	18
From making heap to first turning	2	2	10	20
From first turn to maturity	10	24	20	44
Transport of product and incorporation	2	4	22	48

Total time for programme is 22 weeks minimum, 48 weeks maximum.

The start of the education stage should be arranged to allow sufficient time for waste collection before the date set for building the compost heap. The process of education includes the training of both the extension staff and the farmers. The training may be done in short intensive courses; these may be better for the extension workers. It can also be done over the course of three or four training sessions during the normal extension visits to the village. Adding together the probable timings for all aspects of the programme it will be seen that the absolute minimum schedule is 22 weeks but that it could take up to 48 weeks.

Before the schedule is finalized it must be set against the local farming calendar. Obviously, intensive training courses would not be appropriate at peak activity times such as rice planting, sugarcane harvesting or maize weeding. Care should be taken to see that the schedule does not clash with local festivals and holidays.

The schedule should not stop with planting but include planned meetings with the farmers for evaluating the usage of composts. Absence of such evaluation and follow-up with farmers has often been the main reason for failure of extension programmes. The outcome of these meetings will indicate a basis for the following year's programme. The final schedule should be understood clearly by the extension, teaching and research staff, and by the farmers, before it is acted on. This is important to ensure timely decisions by farmers and effective coordination between staff.

8.4 TRAINING THE EXTENSION WORKER

The extension worker must understand thoroughly four aspects of composting:

i. The role of composting in maintaining soil fertility, its place in the local farming system and the probable economics of its use;

ii. techniques of compost heap construction;

iii. appropriate field scale trials of compost use;

iv. how to teach farmers.

He must receive, therefore, a phased programme of training which will give him an overview of the subject, sufficient detailed knowledge to deal with probable questions, a very clear idea of what to teach on each village visit and practice in how to teach. Confidence in himself is vital, but equally important is his willingness to accept the limits of his own knowledge and readiness to refer problems back to his trainers. Farmers will soon learn to appreciate this and will not look upon it as failure.

It is suggested that the extension worker should be trained in the relevant contents of Chapters 1 to 8 of this manual and that each stage be timed to precede by a maximum of two weeks the date of his visit to the village for that topic. The degree of detail to be used will depend on the level of knowledge of the farmers and the skill of the workers. Too much detail will probably do more harm than too little. Training must always include detailed consideration of how to present material to farmers and how to prepare answers to problems raised by farmers. Teaching examples are given in the next section while possible economics are outlined in Chapter 7.

8.5 FARMER TRAINING

The following teaching aids are suggestions only and are not intended to be an exhaustive list. Ideally, each trainer should develop his own aids and the closer they are to the farmers' own situation the more effective they will be. The exercise of making the aids helps the trainer to discover gaps in his own knowledge and develops his own thinking.

Extension experience suggests that farmers are aware of many of their problems and the constraints to their crop yields. The extension process thus becomes one of using existing local knowledge to enable the farmers to define their problems clearly and in a rational order, whereupon they themselves will begin to suggest steps for solving their problems. Hence the extension worker asks questions to find out local knowledge and then fills in the gaps. By doing so, he himself gains many new insights and can relay genuine problems to his colleagues.

8.5.1 Preparatory Education for Farmers

This will deal with the present local agricultural position and should find the answers to three main questions:

- What are the current agricultural practices and crop yields? These can be recorded.

- What crop yields are achievable? The best local results should be noted and an idea of maximum potential yields suggested.

- What are the reasons for the differences?

Reasons for the shortfall between achieved and achievable yields will normally fall into one or more of the following categories:

i. insufficient rainfall. This may be linked to the availability and adequacy of irrigation facilities;

ii. timeliness and effectiveness of agricultural operations. Included in this heading are such items as equipment, labour availability and cash flow;

iii. pest problems. Normally these include insects and diseases. Weeds are generally considered under timeliness;

iv. soil fertility. This is a key area of extension work into which composting logically fits;

v. seed availability, quality and potential, and the acceptability of the produce;

vi. economics.

There are definite interactions between these reasons. Thus a farmer may be unwilling to invest in mineral fertilizer because of unreliable rains and an uncertain crop response; alternatively, the crop variety which responds well to fertilizer may have a low market price. An effective means of listing these constraints is to display a representative object as each constraint is suggested by farmers. An appropriate set would be:

Rain - a pot of water.

Timeliness - a local plough, weeds, a clock.

Pests - diseased crops, pesticide.

Soil fertility - some soil.

Seed - a sample of seed.

Economics - local money.

When an exhaustive list of constraints have been drawn up, farmers can be asked to arrange them into two groups. The first group consists of those within their control and the second of those beyond their control. Careful questioning of the farmers will normally lead to a realization that no constraint is completely beyond their control. Thus, although they cannot control rainfall, they can exercise a lot of control over soil moisture. Their crops may not always escape pest attack but by using resistant varieties, practising rotations and spraying promptly they can minimize loss. A consideration of the relationships between constraints and the control that farmers can establish provides a good understanding of the relative importance of each constraint and a good basis for further teaching on soil fertility, or indeed on any other constraint.

8.5.2 Education on Soil Fertility

The main questions to pose to farmers on this subject are:

- What is soil fertility?

- How is fertility broken down?

- How is fertility regenerated and improved?

The factual answers to these questions are found in Chapters 1 and 5. Some teaching suggestions are set out below as a stimulus to thinking.

i. **Soil fertility**

Composition. Put 60 mm of soil in a tall transparent glass jar. Fill the jar almost full with water and shake it vigorously. Allow several hours for the soil to settle. Individual components will settle out separately. This can be done with all the soil types in the village.

Appearance. Dig pits 450 mm square and up to 450 mm deep in the best arable, worst arable and the grazing lands of the village. Compare the exposed surfaces. Note any differences in composition or structure.

Organic matter. Pour some hydrogen peroxide on to samples of soil, organic matter and river sand. Assess the amount of reaction.

Discuss the findings of all the tests and build up a picture of the soil as a living world. Ask for examples of local soil animals such as termites, beetles and worms.

ii. **Fertility breakdown**

Study the organic and nutrient cycles shown in Figures 1, 2, 4 and 6. Look at erosion problems at various sites in the village. Compare the soils of these areas with those of the best arable land (Figure 41). Work out reasons for the differences.

Ask the farmers to give the history of their lands for as long as they can remember. Are yields increasing, stable or declining? What agricultural practices have changed?

Ask if rock outcrops are becoming bigger and if tree roots are becoming more exposed. If the answers are yes, find out why.

Find out if streams flow for as long as they used to and if tanks (reservoirs) irrigate as much land as previously? Do streams flood more frequently now? Do soils dry out more quickly than they used to? Ask if rainfall is increasing or decreasing. Are there reliable rainfall data available?

If possible, carry out simple colour tests on soil samples for nutrient availability. Test the pH of saline areas with universal test papers. Cheap test kits may be available from the government advisory service or agricultural traders and suppliers.

Discuss all the results and answers in relation to erosion, structure, soil moisture characteristics and nutrient levels.

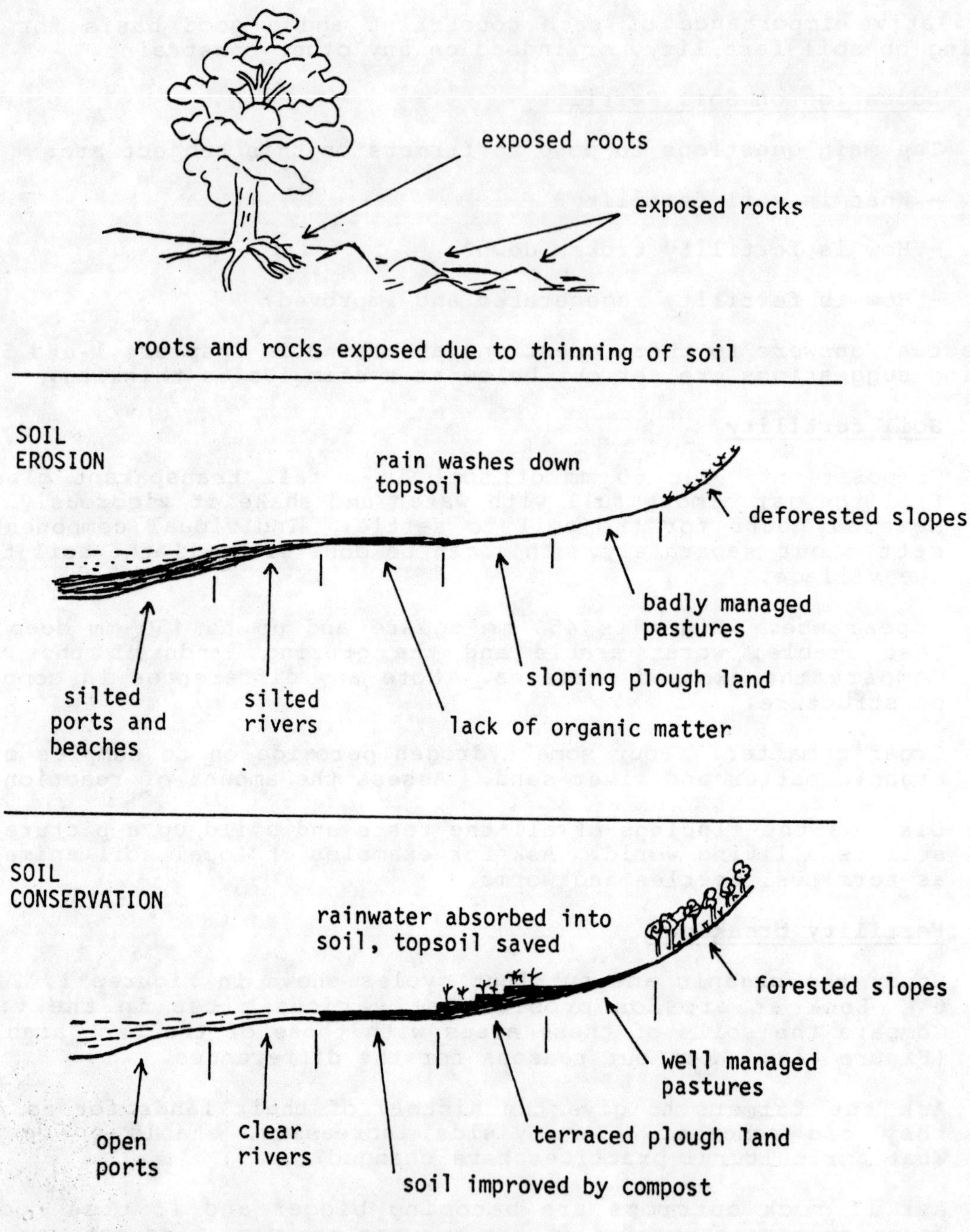

Figure 41 Soil fertility breakdown

iii. **Fertility regeneration and improvement**

It will be necessary to deal with erosion control methods. This is a complete training process in itself and reference should be made to FAO Soils Bulletins 30, 33, 44, 52 and 54 (FAO 1976, 1977, 1979, 1983 and 1984). Such control is essential in many tropical areas because soil fertility cannot be maintained if the soil is not conserved. A consideration of the re-establishment of tree cover will also be necessary.

Consider the traditional usage of manures. Add some compost to a sample of sand and then wet it. Roll the damp mixture between the hands and compare it with rolling damp sand. Has it begun to stick together on mixing? Repeat the earlier exercise of using hydrogen peroxide to test for organic matter in soils of differing fertility.

Fill a dry earthen pot with dry sand. Fill a second pot of the same size with a mixture of 4 parts dry sand and one part dry compost. Fill a third pot to a depth of 25 mm from the top with dry sand and cover it with a mulch of dry compost 20 mm thick. Add an equal quantity of water to each pot and keep them in the shade. Test the dampness every three to four days and observe the different drying times. This demonstration can be very effective if adequate pre-testing is done to determine the ideal quantities of sand and water to use to prevent total drying of all samples before the next training session.

Ask about crop rotation practices and traditional intercropping. Show samples of roots of different crops such as maize and pigeon pea. How deep do the roots go? Look at the organic wastes which are currently recycled, especially those which are thrown casually into pits without mixing, watering or turning. Do some colour tests and water solubility tests on various locally available mineral fertilizers, soils and composts. Draw up a chart showing the benefits and complementary roles of organic manures and mineral fertilizers. Suggestions can be found in Table 19.

These considerations of the basis of soil fertility management will help the farmer to think about his knowledge and fill in major gaps; they will also indicate whether composting is likely to be useful.

8.5.3 Feasibility of Composting

Three major questions must be considered. It is important that all decisions arising from this consideration are made by the farmers. It is also important that the decisions are based on local conditions.

- What wastes are available?

- How much will composting cost?

- How will composting fit into the farming year?

The question of waste availability can be answered by doing an inventory of existing local wastes and then thinking about ways of producing more material. Examples shown in Figure 42 include:

i. alternatives to burning dung for fuel, such as biogas generation, firewood production and husk burning;

ii. planting trees like *Gliricidia* species which will generate green material for composting;

iii. improvement of local fodder production using legumes like *Stylosanthes* species and *Leucaena* species (the 'ipil ipil' tree). This can release low nutritional value cereal by-products for composting;

iv. the possibility of community level sanitation schemes which will facilitate hygienic handling of night soil for composting.

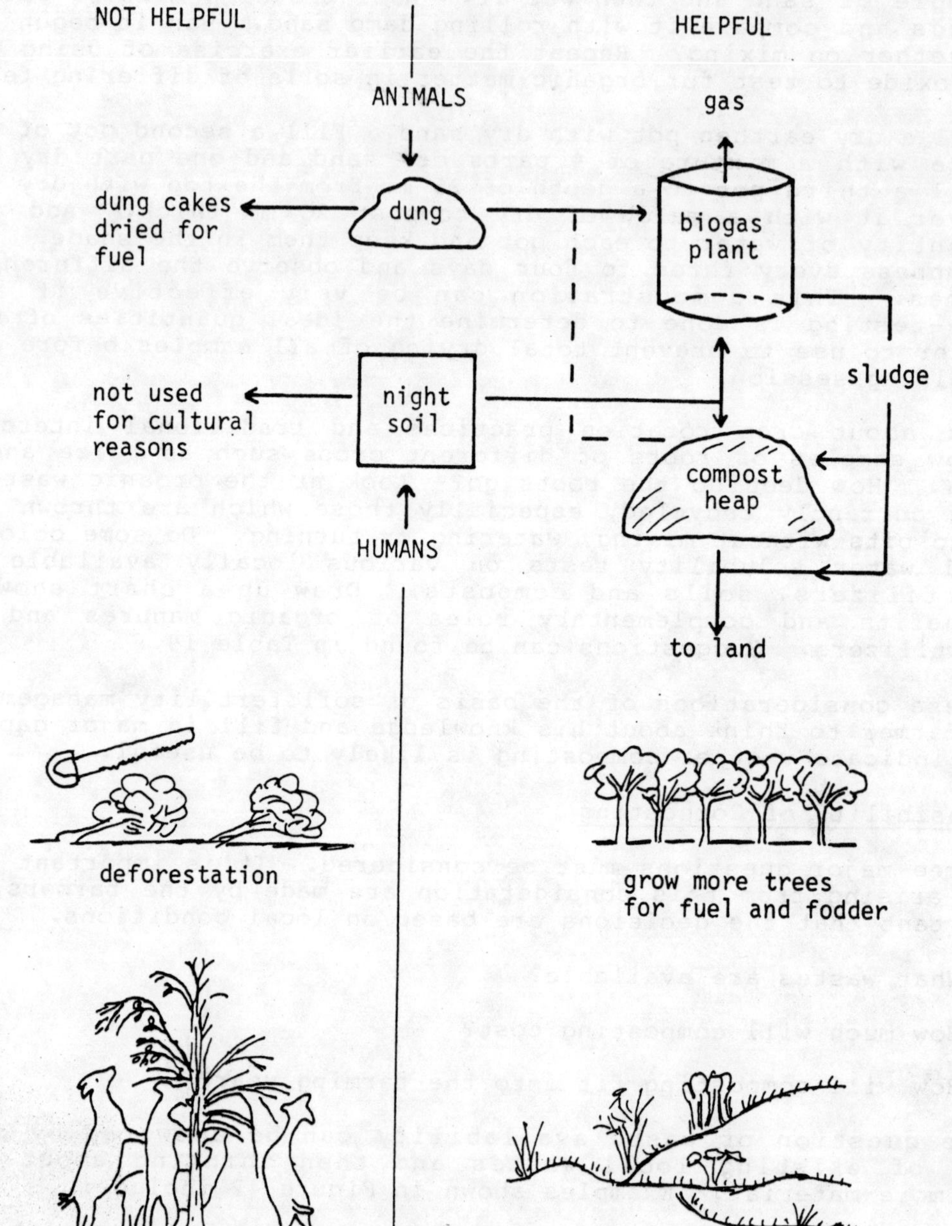

Figure 42 Increasing the availability of organic wastes

Whichever organic wastes are ultimately used for composting, two important facts must be borne in mind. Firstly, when a harvest is removed from a field, loss of nutrients and organic matter occurs and can only be made good from another source. The greater the yield of the crop, the greater will be the loss from that field, and hence the greater the need for replacement. Secondly, much of the material returned will come from a different area and this process will involve a mining or lowering of nutrient and organic matter levels of that soil unless tree crops are bringing new nutrients into it from deeper soil layers.

The only other source of nutrient recovery or recycling is nitrogen fixation by bacteria such as Azotobacter or by Azolla. Composting does not generate new materials but re-organizes and recycles harvested ones.

The question of the cost of composting falls into two sections. One is the cost of collecting and processing the wastes; this may have to be considered at the community and individual level. Some guidelines are given in Chapter 7. New equipment may be required but the cost can almost certainly be at least partly offset against its use for other activities. The other costs to be considered are the long-term ones associated with a decline in soil fertility, lowered yields, loss of crop lands, and the silting-up of reservoir tanks and waterways, Figure 39. For waterways, costs are not only in terms of repair but also of lost irrigation to crops.

The final feasibility question of dovetailing the composting process into the farming year can only be answered against the local background. The time schedule given in Table 28 can be used as a general guide. Some questions such as alternative night soil handling and biogas generation may have to be left for following seasons. It is very important, however, for farmers to realise how all factors of crop and animal production, fodder, fuel and soil fertility are integrated into a complete programme and how shifts in one factor may affect one or more of the other factors.

Farmers' involvement, enthusiasm and understanding can be greatly increased if they are given the responsibility for calculating waste availabilities, estimating time schedules and labour needs. The final decisions arrived at by farmers should be made against the background that any system of soil management which ignores the importance of organic matter in building up soil fertility will not be a progressive or stable one. The question is not whether or not to return organic matter, but rather how to maximize the return and how best to complement it with mineral fertilizer.

To ensure the best decisions much of the education process in each village will need to use as visual aids the fields, waste lands, grazing lands, trees, waterways, kitchens and latrines of the village. Effective charts can be made to link these and the farmers and their families together in the overall economy.

When farmers are satisifed about the usefulness of composting as a soil management tool, the extension worker can then assist them with detailed planning of the logistics of heap or pit siting and preparation, water supply and transport of materials. Collection of organic wastes can then proceed. The next extension task will be to assist with making the compost heaps.

8.5.4 Compost Heap Construction

An extension visit should be planned to coincide with the start of heap construction. The extension worker should have gained practical experience elsewhere of waste preparation, blending, moisture adjustment and aeration before he begins the first heap construction in a new village. This exercise is entirely practical and no visual aids will be needed. Details of equipment and heap assembly are given in Chapter 4.

When he is satisfied that the farmers have a good grasp of the practical points of heap construction, the worker can leave them to proceed. He should time another visit for the first turning of the heap in

14 days time so that he can advise on moisture and aeration adjustment if this process is necessary. If turning is not required, the temperature and moisture tests described in Chapter 4 should be carried out. If more than one turning is planned then dates for these can be arranged, although the presence of the extension worker should not be necessary at these turnings. A further check six weeks before expected use of the compost should be sufficient to assess progress and whether a last turning is advisable.

8.5.5 Monitoring the Use of Compost

After heap construction attention can be given to trials to test different methods of compost use. At this time farmers' motivation will be high; they will be on their way to having a usable product and it will be close enough to planting time for all concerned to agree on the objectives of the trials and the necessary recording of the results. The selection of the farmers to carry out the trials is best left to the farmers themselves. In this way enthusiastic farmers who are held in respect and are likely to be copied will be chosen. Each trial should be of one factor only and should be done on a small portion of the farmer's plot. Other producers can replicate this trial. Other trials can be carried out with different farmers in the same season. Experience shows that a village level extension worker cannot be reasonably expected to supervise many more than 4 trials with 3 replicates of each.

The main questions to be answered are likely to be:

- Did using compost help with soil moisture management?

- Did it improve crop yields?

- Is there any difference in the soil following use of compost?

- Did using compost reduce soil erosion?

- Was there any effect on weeds and pests?

- Was it profitable?

It may be necessary to repeat trials over several seasons to answer some of these questions. A change of crop need not interfere with the trials. Some trials which are relevant to normal small farmer practice are listed below but many others are possible:

i. compost plus mineral fertilizer compared with mineral fertilizers alone. As far as possible the total nitrogen application should be the same for each treatment;

ii. compost application only compared to mineral fertilization only. Again nitrogen application should be balanced if possible;

iii. compost mulch on the surface compared with compost ploughed into 100 mm depth of soil;

iv. compost spread evenly over the whole plot compared with compost distributed along the rows.

These trials are field level operational research and hence the precision of research station trials is not necessary. Some background information and some reasonably accurate measurements will be needed if the trials are to be interpreted meaningfully in comparison with each other and between different seasons.

Background data which are required are:

- Rainfall, preferably on a daily basis. Where recording schemes are not in existence it may be useful to start. During long dry spells extension visits should be made to evaluate water-holding capacity of the soil and the condition of the crops.

- Sowing dates.

- Weed populations and weeding dates.

- Insect and disease problems; dates and control methods.

- Nutrient deficiency problems if any.

Accurate measurements which are needed are:

- The N, P and K values of the compost if possible. This is a laboratory task.

- Application rates for composts and application rates and compositions of mineral fertilizers.

- Dates of application of compost and fertilizers.

- Yields of crops grown

In some cases, subject matter specialists may be interested in obtaining other data. Special arrangements may then be necessary.

It is important to ascertain the farmers' reactions during the growing season, to point out differences and to arrange for evaluations after harvest. This whole process will lead to the isolation of positive points for future practice and indicate areas for further assessment of negative points. It is very important that timely and effective management advice is given at critical growth times such as weeding, top dressing and when pests become a problem. This will ensure that differences between treatments will not be due to inefficient application of the treatments.

The culmination of the extension programme in the post harvest evaluation will be the starting point for the next season's extension work.

CHAPTER 9

CONCLUSIONS

In many areas of the tropical and subtropical world there is a need to increase the production of food, fodder, fibre and fuel to meet the needs of rapidly increasing populations. There is also a pressing need to reduce the risks and fluctuations in agricultural production from year to year. Soil fertility in some of these areas is declining as a result of attempts to increase output without sufficient attention to the prevention of erosion and the maintenance or improvement of soil fertility. Deforestation of vast areas without replanting has caused widespread erosion in many cases, leading to a major lowering of the productive capacity of the land.

It is widely accepted that the main factor in soil fertility is the level of soil organic matter. Organic matter improves the soil structure and thus enables the soil to resist erosion, to hold more water without waterlogging, to remain moist for a longer time during dry spells and to hold a greater reserve of plant nutrients. It helps to prevent the build-up of extremes of soil acidity or alkalinity. Many of its beneficial effects are due to the stimulus that it gives to soil micro-organisms and small soil animals. In tropical soils high temperatures lead to very rapid oxidation of soil organic matter; hence the need to apply large quantities of organic matter to the soil as often as possible is more urgent than for temperate climates. This can be done by returning to the soil almost all crop, animal and human wastes and by growing trees and shrubs which will provide green material for composting and mulching. Such trees and shrubs can also bring up nutrients from deep subsoil layers.

Composting offers a means of processing a very wide range of organic wastes to form humus which is the most stable form of soil organic matter suitable for incorporation into the soil. Composting has the further benefits of minimizing the spread of disease-causing organisms, reducing waste volume and hence transport costs, and of creating productive opportunities for labour. The use of compost may enhance the response of crops to mineral fertilizers. In many cases compost will prove to provide the cheapest forms of nitrogen, phosphorus and potassium. Composts also provide a wide range of trace elements. There is increasing evidence that composts can be of high value in fish farming. Care must be taken in the composting process to ensure destruction of pathogens and weeds. For certain crops the use of immature composts may be harmful. Composting should not be seen as an alternative to the use of mineral fertilizers; the two types of manuring have different and complementary functions.

The technology of composting is well understood and can be adapted to suit the needs of the processor and local conditions. Thus systems are available at varying levels of cost and complexity to enable public authorities to process urban domestic and industrial wastes in quantities up to 1000 tonnes per day. Different systems using the same basic principles are suitable for farm scale composting. The equipment required for farm systems is simple and can be used in other farming activities and in rural transport and building trades.

Production costs of compost will depend largely on local labour and transport rates. The competitiveness of the compost product in the farm economy will be related to the cost of alternative mineral fertilizers. Comparison of costs on the basis of a single application can be very misleading; farming systems which ignore the importance of soil organic matter levels will, in the long term, lead to the very expensive costs of increased maintenance, and perhaps obsolesence, of irrigation channels and dams, caused by erosion and silting up. In addition there will be the lowering of production levels and, in some cases, actual loss of productive land.

For composting to become more widespread in tropical and subtropical agriculture there is a need to educate all sectors of the community in integrated methods of organic waste disposal and production of food, fodder, fuel and fibre. Local authorities need to consider the cost of handling wastes and the usefulness of the end product. The development of imaginative, responsive and well-organized agricultural extension services will involve farmers fully in responsible decision making and encourage them to adopt agricultural practices which will improve and stabilize production and at the same time conserve and improve the soil which is their main productive asset.

APPENDIX 1

CONVERSION FACTORS AND ABBREVIATIONS

i. Length, area, volume, mass and dressing rates.

1 kilometre (km)	=	1000 metres (m)
1 metre (m)	=	100 centimetres (cm)
1 metre (m)	=	1000 millimetres (mm)
1 metre (m)	=	1 000 000 microns (μm)
1 hectare (ha)	=	10 000 square metres (m^2)
1 cubic metre (m^3)	=	1000 litres (l)
1 tonne (t)	=	1000 kilogrammes (kg)
1 kilogramme (kg)	=	1000 grammes (g)
1 gramme (g)	=	1000 milligrammes (mg)
1 tonne per hectare (t/ha)	=	1 kilogramme per 10 square metres (1 kg/10 m^2)

ii. SI (metric) unit conversions to British units.

1 metre	=	39.4 inches
1 centimetre	=	0.394 inches
1 millimetre	=	0.039 inches
1 hectare	=	2.47 acres
1 litre	=	0.22 UK gallons
1 tonne	=	0.982 tons
1 tonne/hectare	=	0.397 tons/acre

iii. Chemical conversions of fertilizer elements P, K and Ca to the oxide form.

P x 2.292 = P_2O_5	P_2O_5 x 0.436 = P
K x 1.205 = K_2O	$K2_O$ x 0.830 = K
Ca x 1.399 = CaO	CaO x 0.715 = Ca

iv. kWh = Kilowatt hour (a unit of power consumption)

APPENDIX 2

BASIS OF NUTRIENT CALCULATIONS

To compare the nutrients supplied by compost and by mineral fertilizers.

Let

N_C be the percentage of a Nutrient in the Compost, from 0.5% N, 0.2% P and 0.3% K.

N_F be the percentage of that nutrient in the Fertilizer, from Table 21.

A be the percentage of that nutrient Available to the crop from the compost in the year of application, from N = 25%, P = 100%, and K = 80%.

Then

1. a. $N_C \times 10$ = weight of nutrient in kg in 1 tonne of compost.

   b. $\frac{N_C \times 10 \times A}{100}$ = weight of nutrient in kg in 1 tonne of compost available to the crop in the year of application.

   $= \frac{N_C A}{10}$

   c. $\frac{N_C A}{10} \times \frac{100}{N_F}$ = weight in kg of mineral fertilizer to supply an equivalent weight of nutrient to the crop as one tonne of compost in the year of application.

   $= \frac{10 N_C A}{N_F}$

2. a. $\frac{N_F}{2}$ = weight in kg of nutrient in 50 kg of mineral fertilizer.

   b. $\frac{N_F/2}{\frac{N_C A}{10}} = \frac{5 N_F}{N_C A}$ = number of tonnes of compost providing an equivalent weight of nutrient to the crop in the year of application as 50 kg of mineral fertilizer.

EXAMPLE 1.

Find the weight of urea supplying the same weight of nitrogen to a crop as 1 tonne of compost in the year of application.

Concentration N_C = 0.5%

Concentration N_F = 46%

Availability A = 25%

From 1.c.

$$\frac{10\ N_C\ A}{N_F} = \frac{10 \times 0.5 \times 25}{46} = \frac{125}{46} = 2.72\ \text{kg}$$

EXAMPLE 2

Find the weight in tonnes of compost supplying an equivalent weight of potassium to a 50 kg dressing of Muriate of Potash.

Concentration N_C = 0.3%

Concentration N_F = 51.5%

Availability A = 80%

From 2.b.

$$\frac{5\ N_F}{N_C\ A} = \frac{5 \times 51.5}{0.3 \times 80} = \frac{257.5}{24} = 10.73\ \text{tonnes}$$

APPENDIX 3

PRACTICAL "DO'S" AND "DONT'S" OF COMPOSTING

DO think of the Compost Heap as a living structure composed of millions of micro-organisms. If you look after their needs they will make good compost for you.

DO use organic materials only, except for suitable additives.

DO shred or chop large pieces of material and tough stalks.

DO use a compost bin for small heaps if possible.

DO provide a proper means of aerating the heap.

DO build a small heap with the maximum amount of material at one time.

DO water the wastes carefully, to the condition of a damp sponge.

DO provide a cover layer for the heap to help retain heat and to protect it from rain and wind.

DO store finished compost under cover to prevent rain leaching out the water-soluble nutrients.

DON'T put crockery, glass, metal or plastic in the heap as they will not decompose.

DON'T add little quantities of wastes to a heap each day as they will not heat up.

DON'T use thick layers of any one material, especially grass cuttings or newspaper, because they will restrict the air flow.

DON'T compress the compost heap by walking on it.

DON'T over-water and make the heap soggy or it will go anaerobic.

DON'T use the product compost until it is mature enough for the task required; this is most important when used sowing small seeds.

APPENDIX 4

BIOGAS PRODUCTION

When organic wastes are digested in warm wet anaerobic conditions one of the major products is methane gas (Plate 30). This gas can be used for fuel purposes in lighting, cooking, and running specially adapted internal combustion engines. It is not possible in this appendix to consider in full detail the design, operation and economics of biogas plants. Suggestions for further reading are given in Appendix 7.

The gases produced and used as fuel are composed mostly of carbon and hydrogen and hence the mineral nutrients of the organic raw materials are retained in the sludge which is recovered (Plates 5 and 31). This is a big advantage compared with burning the raw materials directly, when all the nitrogen will be lost. As digestion takes place under carefully controlled conditions in an enclosed vessel there is no problem of odours or flies. The process achieves fair control over pathogens but is not totally effective in killing all types of disease organisms found in night soil. This is especially true for some intestinal worm eggs and certain types of Salmonella which can cause serious forms of food poisoning. However, the process greatly reduces the risk of spread of these diseases compared to the direct use of the raw wastes on the land.

Gas can be produced from a wide range of raw materials including wastes from crops, humans, livestock, poultry and food processing. Night soil can be piped from the latrine direct to the digester; this removes all the hygiene and social taboo problems of night soil collection and handling. In some areas there may be an initial social reaction against using gas from human waste but this is not a long-term problem. Some degree of waste processing and control of moisture content may be necessary before the wastes are introduced to the digester; this is normally easier for livestock manures, hence these tend to be the main raw material.

Care must be taken over storing the sludge remaining after digestion to prevent loss of nutrients by leaching or as ammonia gas. The precautions are the same as those required for the storage of organic wastes before composting, as described in Chapter 3.

There is a very wide range of biogas generating plants, and stoves and lamps suitable for burning biogas. Large numbers of plants have been installed and are now running successfully in China, India and other countries in Asia, in Africa and South America. Environmental temperatures in these countries are normally sufficient to allow digestion to take place without the need for supplementary heating. This means that simple engineering designs are successful and as a result biogas plants are much more likely to be economic in rural areas. Even so the initial capital cost is likely to be relatively high, especially if steel is used for the gas storage chamber. In most parts of India the initial capital cost of a 3 m^3 plant, suitable for a family of five persons, is equivalent to the income from 300 labour-days.

This financial disadvantage can be overcome to some extent by building community plants but often the associated social organization problems make such schemes difficult to administer. A further disadvantage is that the plants require management knowledge to control the solids content of raw materials and to ensure a continuous and efficient reaction. Management is normally exercised by someone other than the housewife who is the most immediate beneficiary of the process. As a result a great number of plants are run at very low efficiency and many have fallen into disrepair. This indicates that the introduction and impact of biogas in the long term has an extension education and social organization demand similar to that for composting.

Plate 30 Biogas unit showing inlet from pig-sty, the digester and the sludge outlet pipe
Source: FAO filmstrip Biogas in China

Plate 31 Removal of biogas sludge from the sludge outlet pipe
Source: FAO filmstrip Biogas in China

There is often an apparent conflict between the advocates of biogas production and those of composting. This is an unfortunate misunderstanding as the two processes are complementary. Thus readily digestible materials such as animal manures can be used to produce methane gas and the resulting sludge can then be aerobically composted with high C/N ratio wastes such as straw. Although the sludge could be used directly on the land, the straw would compost less satisfactorily without the sludge. By combining biogas production and composting much of the carbon and hydrogen in the wastes will be used as fuel and the mineral nutrients will be available for crop growth. There will be almost total control of pathogens and the resulting compost product will be in a form which is suited to the improvement of soil structure with all the resulting agricultural benefits. Biogas production can also go a long way to reducing the demand for wood for fuel which has led to so much deforestation.

One economic and social aspect of biogas production is beginning to appear. The start of a new digester requires quite a substantial volume of material, possibly 30 days supply at the normal input rate. The person owning such a digester will buy or trade for these wastes from his neighbours, thus reducing their small holdings of wastes and setting a market value on them. Likewise, the digester owner wishing to increase the regular output of gas will buy waste materials. This practice is perfectly satisfactory provided that the biogas sludge is returned to the neighbouring smallholdings to maintain their supply of plant nutrients.

APPENDIX 5

GLOSSARY

ABSORBED: taken in, sucked in.

ADSORBED: attached to the surface.

AEROBIC: in the presence of air or oxygen.

AMBIENT: surrounding air.

AMMONIA: a gas composed of nitrogen and hydrogen which can be produced during the breakdown of organic wastes.

ANAEROBIC: in the absence of air or oxygen.

AVAILABLE: capable of being used.

BIENNIAL: plant that requires two years to complete its life cycle, from seed germination to seed production and death.

BULK DENSITY: the ratio of the mass of oven dried soil to its bulk volume.

BUND: a low bank between fields; the main wall of a tank or reservoir.

CAPILLARY: a narrow passage through the soil along which air or water can move.

CARBON DIOXIDE: a gas composed of carbon and nitrogen which is produced during the breakdown of organic wastes.

CATALYST: a substance that will speed up a chemical reaction without itself being used up in the reaction.

CATION OR BASE EXCHANGE CAPACITY (CEC): the total quantity of cations which a soil can adsorb by cation exchange, usually expressed in milliequivalents per 100 grams.

CHELATE: the binding of a metal into a complex organic molecule.

CHEMICAL BASES: substances which combine with acids to form chemical salts.

COLLOIDAL: made up of extremely small particles often showing a considerable attraction for water.

COMPANION PLANTING: planting two crops together for each others' benefit. If maize and beans are planted together maize will give support to the bean climber and the bean will fix atmospheric nitrogen and provide a fertile soil for the maize.

CONTOUR: an imaginary line linking points of equal elevation. Contour ploughing is ploughing along the contour line so that all points of the plough furrow are at the same level.

CONVECTION: upward movement of warm air causing cold air to be drawn in from below.

DETRITUS: organic debris from decomposing plants and animals.

ECOLOGY: the study of habitats and lives of plants and animals in relation to environment.

ENVIRONMENT: the surroundings.

ENZYME: a biological catalyst.

EQUILIBRIUM: a state of balance.

EROSION: the gradual disintegration of the soil surface by chemical or physical weathering.

ESSENTIAL FATTY ACIDS: polyunsaturated fatty acids which cannot be formed in the body and must be taken in the diet.

EVAPORATION: turn into vapour.

HEAP: a mass of composting organic material. The heap may be built on the ground when it is referred to as a STACK or may be built partly below ground in a PIT.

HUMUS: complex and stable organic matter resulting from the decomposition of plant and animal material.

HUMIFICATION: the process by which humus is formed.

HYPHAE: fungal filaments which are also called mycelia.

INCORPORATION: to mix in with the soil.

INTERCROPPING: growing two or more crops at the same time.

IRRIGATION: controlled application of water.

KWASHIORKOR: nutritional disorder of children when the diet is deficient in protein; common where maize is the staple food.

LABOUR-DAYS: a day's work by a man or woman.

LEACHING: the washing of organic or mineral substances into the lower soil layer by water from rain or irrigation.

LIGNIN: complex aromatic compound which is present in the walls of some plant cells making them strong and rigid. It forms 20-30 percent of the wood of trees.

MATURE: when composting material has broken down completely to form humus.

MESOPHILIC: ability to grow at moderate temperatures.

MICROCLIMATE: the climate of a specific small area, which often differs from the main climate pattern.

MINERAL: any substance which is not animal or vegetable in origin. Animals and vegetables may have minerals in them.

MINERALIZATION: the breakdown of animal or vegetable matter to release the minerals it contains.

MULCH: a covering over the soil.

MYCELIA: fungal filaments or hyphae.

MYCORRHIZA: the association between fungi and plant roots which improves the absorption of nutrients by the plant from the soil.

NITRIFICATION: conversion by soil bacteria of organic compounds of nitrogen, unavailable in green plants, into available nitrate.

NUTRIENT: serving as food for humans, animals or plants.

ORGANIC: material occurring in plants and animals.

OXIDATION: combining with oxygen. Organic matter is mainly oxidized to carbon dioxide and water.

OXYGEN DEMAND: the oxygen used by organic matter during the course of its decomposition.

PARAMETER: a factor or condition controlling a reaction, such as aeration and moisture content in composting.

PATHOGEN: a disease-causing organism.

PERENNIAL: plant that continues its growth from year to year.

PERMEABILITY: admitting the passage of air or water.

PERCOLATION: water making its way down through pores in the soil.

pH or REACTION: a measure of acidity or alkalinity. For soils a value of 7.0 is neutral; the lower the figure the greater the acidity and the higher the figure the greater the alkalinity.

PHOTOSYNTHESIS: the process by which plants are able to use the energy of sunlight to create food (carbohydrates) from carbon dioxide in the atmosphere and water absorbed.

PLANKTON: small organisms, plants or animals, in seas, lakes or ponds which float or drift almost passively.

POLYMER: a chemical compound built up by the joining together of a large number of identical simpler compounds.

PROTEIN: very complex organic compound, composed of numerous amino-acids.

PSYCHROPHILIC: capable of growing at low temperatures.

PUTRESCIBLE: material which will rot very quickly, giving off bad odours.

RUN-OFF: the water that does not get absorbed in the soil and which flows away.

SANITATION: protection of health against dirt and infection; gathering up human and animal wastes in a hygienic manner.

SAVANNA: mainly flat grassland area in the tropics with very few trees and shrubs.

SEEPAGE: a leakage of liquid out into the soil from compost heaps or latrines.

SEWAGE SLUDGE: the deposit left behind after the treatment of sewage. Sewage is the water-borne waste from houses and towns carried away in underground pipes and drains.

SILVICULTURE: growing trees.

SOLE CROP: growing a single crop.

SPORES: single-celled or several-celled reproductive bodies that become detached from the parent and give rise either directly or indirectly to new individuals. They are produced in large numbers and occur particularly in the lower plants such as bacteria and fungi.

SYMBIOTIC: association of dissimilar organisms to their mutual advantage.

THERMOPHILIC: capable of growing at high temperatures.

TILAPIA: fresh water fish.

TRANSPIRATION: the loss of water vapour from land plants through thin leaves.

WATERLOGGING: the saturation of the soil in the root zone of crops.

WILTING: condition of plants in which cells lose their turgidity; leaves and young stems droop due to an excess of water loss by transpiration over absorption of water through the root system.

WINDROW: long compost stack above ground.

XEROPHTHALMIA: an eye condition associated with lack of Vitamin A which may lead to blindness.

APPENDIX 6

REFERENCES

These references are given in full to help their retrieval from library systems.

Acharya C.N. 1950 Preparation of compost manure from town wastes. Indian Council of Agricultural Research, Bulletin No. 60. Indian Ministry of Agriculture, Calcutta.

Banerjee R.K. and Srinivasan K.V. 1983 Composted urban refuse and primary sewage sludge as a fish-pond manure. Agricultural Wastes 7(4): 209-219.

Benor D. and Baxter M. 1984 Training and visit extension. International Bank for Reconstruction and Development/The World Bank, 1818 H Street N.W., Washington D.C. 20433, USA.

Bertoldi M. de, Vallini G., Pera A. and Zucconi F. 1982 Comparison of three windrow compost systems. Biocycle 23(2): 45-50.

Bertoldi M. de, Vallini G. and Pera A. 1983 The biology of composting: a review. Waste Management and Research 1: 157-176.

Biddlestone A.J., Ball D. and Gray K.R. 1981 Composting and urban waste recycling. In: Advances in food-producing systems for arid and semi-arid lands. Manassah, J.T. and Briskey, E.J. (ed). Part A, 191-224. Academic Press, London.

Dalzell H.W. 1981 An appropriate technology for Indian agriculture. In: Biological husbandry. Stonehouse, B. (ed). 205-217. Butterworths, London.

Dhar N.R., Bose S.M. and Gaur A.C. 1955 Influence of calcium phosphate in composting of organic matter. Proceedings of the National Academy of Sciences, India 24A: 473-488.

Douglas J.S. and Hart R.A. de J. 1976 Forest farming. Watkins, London. 197p.

Edwards P., Polprasert C., Pacharaprakiti C., Rajput V.S. and Suthirawut S. 1983 Compost as fish feed, a practical application of detritivory for the cultivation of tilapia. Agriculture 32: 409-413.

FAO. 1968 Food composition table for use in Africa. FAO, Rome, 306p.

FAO. 1975 Organic materials as fertilizers. Soils Bulletin No. 27. FAO, Rome. 394p.

FAO. 1978a Organic materials and soil productivity. Soils Bulletin No. 35. FAO, Rome. 119p.

FAO. 1978b Organic recycling in Asia. Soils Bulletin No. 36. FAO, Rome. 417p.

FAO. 1978c China: recycling of organic wastes in agriculture. Soils Bulletin No. 40. FAO, Rome. 107p.

FAO. 1980 Organic recycling in Africa. Soils Bulletin No. 43. FAO, Rome. 304p.

FAO. Organic materials and recycling in the Near East. Soils Bulletin No.
1982 45. FAO, Rome. 280p.

Finstein M.S., Lin K.W.R. and Fischler G.E. Sludge composting and utilization: review of the literature on temperature inactivation of pathogens. New Jersey Agricultural Experimental Station, Cook College, Rutgers University, New Brunswick, New Jersey 08903, USA. 29p.
1982

Flintoff F. Management of solid wastes in developing countries. WHO Regional publications, South-East Asia Series No. 1. World Health Organization, New Delhi. 244p.
1976

Gaur A.C. Recycling and utilization of organic wastes as fertilizers. Proceedings of the FAO/India/Norway Seminar on the development of the complementary use of mineral fertilizers and organic materials in India. Technical Bulletin (Agricultural) No. 32. Indian Council of Agricultural Research. 109-126.
1978

Golueke C.G. Epidemiological aspects of sludge handling and management. Biocycle 24(4): 50-58.
1983

Gotaas H.B. Composting - sanitary disposal and reclamation of organic wastes. WHO Monograph Series No. 31. World Health Organization, Geneva. 205p.
1956

Gray K.R. and Biddlestone A.J. Agricultural use of composted town refuse. In: Inorganic pollution and agriculture. Ministry of Agriculture, Fisheries and Food, Reference Book 326. HMSO, London.
1980

Grubben G.J.H. Town refuse for vegetable production. In: Solid waste disposal and utilization in developing countries. Schelhaas, R.M. (ed). Bulletin 310: 86-98. Department of Agricultural Research, Royal Tropical Institute, Amsterdam.
1982

Higgins A.J. Technical issues involving sludge and compost use. Biocycle 24(1): 40-44.
1983

Howard A. and Wad Y.D. The waste products of agriculture, their utilization as humus. Oxford University Press, London. 167p.
1931

Moss R.P. Organic matter cycles in tropical soils and husbandry systems. In: Biological Husbandry. Stonehouse, B. (ed). 39-66. Butterworths, London.
1981

Pacey A. Gardening for better nutrition. Oxfam and Intermediate Technology Publications, London. 64p.
1978

Schelhaas R.M. (ed). Solid waste disposal and utilization in developing countries. Bulletin 310. Department of Agricultural Research, Royal Tropical Institute, Amsterdam. 197p.
1982

Shuval H.I., Gunnerson C.G. and Julius D. Nightsoil composting. P.U. Report No. RES 12. International Bank for Reconstruction and Development/The World Bank, 1818 H Street N.W., Washington D.C. 20433, USA. 46p.
1978

Vuren J.P.J. van Soil fertility and sewage. Faber and Faber, London, 236p.
1949

WHO Handbook on human nutrition requirements. WHO Monograph Series No.61.
1974 World Health Organization, Geneva.

Wilson G.F. and Kang B.T. Developing stable and productive biological 1981 cropping systems for the humid tropics. In: Biological husbandry. Stonehouse, B. (ed). 193-203. Butterworths, London.

APPENDIX 7

FURTHER READING, FAO FILMSTRIPS ABOUT COMPOSTING AND USEFUL ADDRESSES

A7.1 Further reading

In addition to the references given in Appendix 6 the following books provide further information on composting and associated subjects.

i. Historical aspects

King F.H. 1949 Farmers of forty centuries. Jonathan Cape, London. 379p.

Howard A. 1943 An agricultural testament. Oxford University Press, London. 253p.

ii. The composting process - general

Golueke C.G. 1972 Composting - a study of the process and its principles. Rodale Press, Emmaus, Pennsylvania 18049, USA. 110p.

Poincelot R.P. 1972 The biochemistry and methodology of composting. Bulletin 727. Connecticut Agricultural Experiment Station, New Haven, Connecticut 06504, USA. 38p.

iii. Composting - biochemistry and microbiology

Alexander M. 1977 Introduction to soil microbiology. John Wiley, New York.

Kononova M.M. 1966 Soil organic matter. Pergamon Press, Oxford.

Waksman S.A. 1938 Humus. Williams and Wilkins, Baltimore, USA.

iv. Composting of night soil, refuse and sewage sludge

Haug R.T. 1980 Compost engineering: principles and practice. Ann Arbour Science, Michigan, USA.

McGarry M.G. and Stainforth J. 1978 Compost, fertilizers and biogas production from human and farm wastes in the People's Republic of China. Bulletin IDRC-TS8e. International Development Research Centre, Ottawa, Canada, 94p.

Singley M.E., Higgins A.J. and Rosengaus M.F. 1982 Sludge composting and utilization: a design and operating manual. New Jersey Agricultural Experiment Station, Cook College, Rutgers University, New Brunswick, New Jersey 08903, USA. 245p.

v. Agricultural aspects

Balfour E.B. 1943 The living soil. Faber and Faber, London. 258p.

FAO. 1976 Prognosis of salinity and alkalinity. Soils Bulletin No. 31. FAO, Rome. 272p.

FAO. Guidelines for prognosis and monitoring of salinity and sodicity. Soils Bulletin No. 39. FAO, Rome (in preparation).

FAO. 1983 Guidelines: land evaluation for rainfed agriculture. Soils Bulletin No. 52. FAO, Rome. 249p.

FAO. 1984a Improved production systems as an alternative to shifting cultivation. Soils Bulletin No. 53. FAO, Rome. 206p.

FAO. 1985a Guidelines: land evaluation for irrigated agriculture. Soils Bulletin No. 55. FAO, Rome. 243p.

FAO. 1985b Water quality for agriculture. Irrigation and Drainage Paper No. 29 Rev. 1. FAO, Rome. 174p.

vi. Soil conservation

FAO. 1976 Soil conservation in developing countries. Soils Bulletin No. 30. FAO, Rome.

FAO. 1977 Soil conservation and management in developing countries. Soils Bulletin No. 33. FAO, Rome.

FAO. 1979 Watershed development with special reference to soil and water conservation. Soils Bulletin No. 44. FAO, Rome.

FAO. 1984b Tillage systems for soil and water conservation. Soils Bulletin No. 54. FAO, Rome.

vii. Desert reclamation

Adams R., Adams M., Willens A. and Willens A. 1978 Dry lands: man and plants. The Architectural Press, London. 152p.

Baker R. St. B. 1966 Sahara conquest. Butterworth Press, London. 186p.

Campbell-Purdie W. and Brockway F. 1967 Woman against the desert. Victor Gallancz, London. 128p.

viii. Human nutrition

Paul A.A. and Southgate D.A.T. 1978 McCance and Widdowson's The composition of food. HMSO, London.

Platt B.S. 1972 Tables of representative values of foods commonly used in tropical countries. Medical Research Council Special Report No. 302. HMSO, London.

ix. Biogas production

FAO. 1979 China: azolla propagation and small-scale biogas technology. Soils Bulletin No. 41. FAO, Rome. 81p.

National Academy of Sciences. 1977 Methane generation from human, animal and agricultural wastes. National Academy of Sciences, 2101 Constitution Avenue, Washington D.C. 20418, USA.

Sathianathan M.A. 1975 Biogas achievements and challenges. Association of Voluntary Agencies for Rural Development (AVARD), New Delhi.

A7.2 FAO Filmstrips about composting and associated topics

The FAO, in cooperation with field projects in several regions of the world, has prepared a number of filmstrips with accompanying commentaries on compost production, soil conservation and agro-forestry. These filmstrips are very useful aids in training and extension work as described in Chapter 8 of this Soils Bulletin. Further information on the filmstrips may be obtained from the Visual Media Section, Information Division, Food and Agriculture Organization of the United Nations, Via delle Terme di Caracalla, 00100 Rome, Italy, or from the Distribution and Sales Section of the FAO at the same address.

Among the titles relevant to this Bulletin are:

i. Compost. This one was produced in Africa and is intended for African audiences.

ii. Compost. This was prepared in northern Thailand and relates to conditions in South East Asia.

iii. A good land forever. This covers several aspects of soil conservation including hillside terracing and was produced in Central America.

iv. Uses of coffee pulp. This filmstrip describes the composting of coffee pulp, and its use for producing biogas and as a cheap feedstuff for some livestock; it was prepared in Central America.

v. Coffee pulp, a good fertilizer. This is a shortened version of filmstrip iv. covering the production and use of compost from coffee pulp.

vi. Biogas in China. This describes the production and use of biogas.

vii. Home-made biogas. This filmstrip describes in detail the construction and use of a biogas digester of a size suitable for one family in China.

A7.3 Useful addresses

i. Action for Food Production (AFPRO), Safdarjung Development, New Delhi, India

ii. Appropriate Technology Development Organization, Planning Commission, Government of Pakistan, Islamabad, Pakistan.

iii. Association of Voluntary Agencies for Rural Development, A-1 Kailash Colony, New Delhi, 110048, India.

iv. Bangladesh Academy for Rural Development, Comilla, Bangladesh.

v. Economic and Social Commission for Asia and the Pacific (ESCAP), Division of Industry, Housing and Technology, United Nations Building, Bangkok 2, Thailand.

vi. International Crops Research Institute for the Semi-Arid Tropics (ICRISAT), Information Officer, Patancheru, Medak District, Andhra Pradesh, India.
This institute has a lot of information on intercropping.

vii. International Institute of Tropical Agriculture, PMB 5320, Ibadan, Nigeria.
This institute has carried out extensive research on mulching for soil conservation.

viii. Intermediate Technology Publications, 9 King Street, London WC2E 8HN, UK.

ix. Intermediate Technology Transport, Home Farm, Ardington, Oxon. OX12 8PN, UK.
This group designed the small farm transport vehicle and cycle trailer unit shown in Figures 28 and 29. They state that it is most important to follow the correct engineering principles to achieve efficient vehicles. They can supply the necessary details.

x. Khadi and Village Industries Commission (Gobar Gas Scheme), Gramodaya, Itla Road, Vile Parle, Bombay 400 056, India.

xi. Volunteers in Technical Assistance (VITA), 3706 Rhode Island Avenue, Mt. Rainier, Maryland 20822, USA.

xii. World Council of Churches, CCPD Technical Services, 150 Route de Ferney, 1211 Geneva 20, Switzerland.

xiii. World Health Organization (WHO), 1211 Geneva 27, Switzerland.

APPENDIX 8

COMMON AND SCIENTIFIC NAMES OF SOME PLANTS MENTIONED IN TEXT

Arecanut:	*Areca catechu* L.
Bamboo:	*Bambusa* spp.
Banana:	*Musa* spp.
Barley:	*Hordeum vulgare* L.
Carrot:	*Daucus carota* L.
Castor:	*Ricinus communis* L.
Chick pea:	*Cicer arietinum* L.
Coconut:	*Cocos nucifera* L.
Cocoyam:	*Colocasia esculenta* (L.) Schott. Also called taro.
Groundnut:	*Arachis hypogaea* L.
Jute:	*Corchorus* spp.
Linseed:	*Linum usitatissimum* L.
Lucerne:	*Medicago sativa* L.
Maize:	*Zea mays* L.
Millet:	*Eleusine* sp. *Panicum* sp. *Pennisetum* sp. *Setaria* sp.
Mung bean:	*Vigna radiata* L.
Pigeon pea:	*Cajanus cajan* L. Also called Red gram, *dal*, *dhal*.
Pumpkin:	*Cucurbita* spp.
Rice:	*Oryza sativa* L.
Russian comfrey:	*Symphytum X uplandicum* Nym.
Sorghum:	*Sorghum bicolor* (L.) Moench.
Sugarcane:	*Saccharum officinarum* L.
Sunflower:	*Helianthus annuus* L.
Sunn hemp:	*Crotalaria juncea* L.
Tobacco:	*Nicotiana tabacum* L.
Tomato:	*Lycopersicon esculentum* Mill.
Water hyacinth:	*Eicchornia crassipes*